Lectures on Digital Design Principles

RIVER PUBLISHERS SERIES ELECTRONIC MATERIALS, CIRCUITS AND DEVICES

Series Editors:

Jan van der Spiegel
University of Pennsylvania, USA

Massimo Alioto
National University of Singapore, Singapore

Kofi Makinwa
Delft University of Technology, The Netherlands

Dennis Sylvester
University of Michigan, USA

Mikael Östling
KTH Stockholm, Sweden

Albert Wang
University of California, Riverside, USA

The "River Publishers Series in Electronic Materials, Circuits and Devices" is a series of comprehensive academic and professional books which focus on theory and applications of advanced electronic materials, circuits and devices. This includes analog and digital integrated circuits, memory technologies, system-on-chip and processor design. Also theory and modeling of devices, performance and reliability of electron and ion integrated circuit devices and interconnects, insulators, metals, organic materials, micro-plasmas, semiconductors, quantum-effect structures, vacuum devices, and emerging materials. The series also includes books on electronic design automation and design methodology, as well as computer aided design tools.

Books published in the series include research monographs, edited volumes, handbooks and textbooks. The books provide professionals, researchers, educators, and advanced students in the field with an invaluable insight into the latest research and developments.

Topics covered in this series include:-

- Analog Integrated Circuits
- Data Converters
- Digital Integrated Circuits
- Electronic Design Automation
- Insulators
- Integrated circuit devices
- Interconnects
- Memory Design
- MEMS
- Nanoelectronics
- Organic materials
- Power ICs
- Processor Architectures
- Quantum-effect structures
- Semiconductors
- Sensors and actuators
- System-on-Chip
- Vacuum devices

For a list of other books in this series, visit www.riverpublishers.com

Lectures on Digital Design Principles

Pinaki Mazumder

University of Michigan, USA

Idongesit E. Ebong

Nixon Peabody, USA

Routledge
Taylor & Francis Group

NEW YORK AND LONDON

Published 2023 by River Publishers
River Publishers
Alsbjergvej 10, 9260 Gistrup, Denmark
www.riverpublishers.com

Distributed exclusively by Routledge
605 Third Avenue, New York, NY 10017, USA
4 Park Square, Milton Park, Abingdon, Oxon OX14 4RN

Lectures on Digital Design Principles / by Pinaki Mazumder, Idongesit E. Ebong.

Routledge is an imprint of the Taylor & Francis Group, an informa business

ISBN 978-87-7022-361-4 (print)
ISBN 978-87-7022-960-9 (paperback)
ISBN 978-10-0092-194-6 (online)
ISBN 978-1-003-42576-2 (ebook master)

Publisher's Note

The pilot version of the current book contains the first ten chapters and it is published to receive feedback from instructors, students and readers, who may submit their comments and errata to the publisher at info@riverpublishers.com

The rest of the twelve chapters will comprise finite state machine design, high level state machine design, register transfer level design, adder design, multiplier and divider design, data path and control logic design, memory design, and a complete processor design, from the specified instruction set architecture.

Due to Pianki Mazumder sustaining severe spinal cord injury after a recent accident, the second part of this book will be delayed by a few months. The pilot version of the second part of the book will be published in a similar manner to the first part to receive feedback from the readers. Subsequently, all twenty two chapters will be combined to create an integrated single book.

Contents

Preface

Thomas Watson, the president of IBM, once boldly predicted: "I think there is a world market for maybe five computers." At that time, the large-size computers such as IBM 704 and IBM 709 were made of bulky and power-hungry vacuum tubes, magnetic core and magnetic tape memories, and CRT displays. Customers for gigantic computers that warranted a separate air-conditioned building were few and far between in the early years of commercial digital computing machines. Almost 30 years later, Ken Olsen, the founder of Digital Equipment Corporation, which manufactured mid-size computers such as PDP-11 and VAX-11, pronounced definitively: "There is no reason anyone would want a computer in their home." In other words, Olsen did not envision the market potential of personal computers like Apple 1 that Steve Jobs and Steve Wozniak had just introduced in the market in 1976. Watson and Olsen were pioneers in the computer industry who had wrought upon the miracle of exponential growth of computing power that has fueled the accelerated economic growth in the last 70 years. Nevertheless, these doyens in computing were dead wrong in their prophecy! Now, 40 years later, more than one billion individuals in the world personally possess at least five computing devices in their notebook, cell phone and smart watch. Globally, more than 100 billion computing chips are now deployed ubiquitously in every walk of the modern world!

This amazing story must inspire you to become a computer hardware engineer who will push the envelope of computing in the 21st Century. Moore's Law in microelectronics that prognosticated the exponential growth of computing power by shrinking geometries of transistors and wires in microprocessor chips, is likely to sustain for a long time and hardware engineers are expected to bring about more astounding inventions that will profoundly impact our lives. The landscape of hardware design is continually unfolding as computers, communications, and consumer electronics are melding together to build the future digital systems equipped with ultra-high-definition cameras and displays. On the other hand, quite logically you may aspire to become a software engineer as alluring job markets in information technology loom upon boundless opportunities for innovations. Nevertheless, you must learn computer hardware very well as in modern portable and energy scavenged systems both hardware and software are optimized holistically as you might have witnessed in smart phones, smart watches and other wearable products.

The purpose of this book that aggregates my lectures in an introductory course for digital logic design is to teach rudimentary principles of digital system design at first and then help you to design, build and test a custom microprocessor using a field programmable gate array (FPGA). There are altogether 21 chapters that I try to cover in my lectures for a one semester 4-credit course supplemented with 6 laboratory experiments to allow students to apply their textbook knowledge and imbibe integrated hands-on training. Students solve small real-word problems and then implement their design on an FPGA prototyping system.

Though I have taught this course rather sporadically ever since I joined the University of Michigan, I did not make any serious efforts to compile my teaching materials to write a book. I used to teach the subject materials by writing on chalkboard from my hand-written notes. The first push to organize my lecture materials came rather unexpectedly, when the National Technology University (NTU), the first accredited "virtual" university, invited me to produce a video book in June 2005 by taping my lectures at the MGM Studio of Disney World in Orlando, Florida. With the support of major technology companies such as IBM, Motorola, and Hewlett-Packard, the on-line university (NTU) was founded in 1984 to deliver academic courses to corporations' training facilities via a unique satellite network. NTU, which has now merged with Walden University, provided an imaginative and unique model of educational integration as a way to award degrees through distance education.

I took this opportunity to convert my hand-written notes into PowerPoint slides that were parceled into 42 lectures for a one-semester on-line course. To produce video streams of my lectures, professional MGM Studio equipment was used in conjunction with Camtasia software running on my tablet PC. Each lecture was nearly 90 minutes and the entire lecture was taped without any retake and stopping. No coughing or sipping a beverage to clear my throat. The streaming technology the equipment used was unforgiving. A few times, someone involuntarily tripped over the long cable running across a big room, thereby compelling me to restart my lecture all over again. A few times, I floundered almost at the end of a lecture and had to restart the entire lecture. The experience was also unique, as I had to sit on a chair inside a small studio cubicle virtually constrained by the viewing span of the camera with bright lights blazing all around. This was a strange teaching experience – sitting like a potted plant and talking to the fixed expressionless camera! As I had not yet published a textbook on digital logic, I had to slightly change my lecture notes to adapt my slides with two textbooks that students could use in conjunction with my lectures and solving practice problems.

Soon after this video lecture production, I contemplated penning down my lectures in the form of a written textbook. But I had to shelve my plan as I went on a US Government assignment to the National Science Foundation at Washington, DC to lead a showcase program on emerging technologies comprising of nanoelectronics, quantum computing and bio-computing. After I returned to the University of Michigan and resumed teaching in 2011, I was asked to teach the logic design course. During the summer of 2011, I started reviewing my videotape to regroup my lectures into different chapters. My co-author, Idongesit Ebong listened very enthusiastically to my lectures, and studied my PowerPoint slides and written notes. He then volunteered to transcribe my lectures and accelerate the writing process. We ended up writing about 250 pages that I adopted in my Fall 2011 class.

The progress became rather slow after that as I became involved in developing a new graduate course on ultra-low-power sub-threshold CMOS[1] systems that is at the heart of design of wearable electronic products. After teaching other regular courses along with this new course several times, I again got opportunity to teach the digital design course in Fall 2015. This enabled me to fully concentrate on writing the textbook. Thanks to Yalcin Yilmaz for helping me in transcribing some chapters.

Finally, computers and digital systems are now ubiquitous, pervading all walks of our daily lives. They constitute the backbone for mission-critical applications in outer space, satellite and wireless communications, military missions in inhospitable terrains, and in underwater explorations. Students must learn the design concepts of digital systems thoroughly, because engineering mistakes are very costly and fatal as we have witnessed in Boeing 737 Max crashes, the Challenger space station disaster, and so on.

Finally, computers and digital systems are now ubiquitous pervading all walks of our daily lives. They constitute the backbone for mission-critical applications in outer space, satellite and wireless communications, military missions in inhospitable terrains, and in underwater explorations. Students must learn the design concepts of digital systems thoroughly,

---

[1]CMOS is acronym for Complementary Metal-Oxide Semiconductor

because engineering mistakes are very costly and fatal as we have witnessed in Boeing 737 Max crashes, the Challenger space station disaster, and so on.

Professional engineers must learn their subjects very thoroughly and acquire disparate skills that will enable them to design their systems after analyzing and critiquing them comprehensively.

Acknowledgement

Though the book is compilation of my lectures and my coauthor has helped me to write these chapters, the structure of the book was conceived from the syllabus of EECS 270: Introduction to Digital Logic Design, a sophomore computer hardware course that is taught by several of my distinguished colleagues at the Computer Science and Engineering Division of EECS Department at the University of Michigan. I must express my unstinted gratitude to them for sharing their homework and exams with me over the course of so many years. I used some of the problems they provided to me as multiple-choice exercise problems at the end of many a chapter.

When I joined the University of Michigan, my senior colleagues John Hayes and Ward Getty were contemplating integrating Mentor Graphics computer-aided design (CAD) tools such as QuickSim and MSPICE into the existing laboratory experiments. They were previously using experiments that required wire wrapping of medium-scale integrated (MSI) chips on breadboards. The breadboard chores for students were pretty laborious and fault-prone because those MSI chips used transistor-transistor logic (TTL) technology that would malfunction if any wire was not firmly connected. When I began teaching the course, they shared with me the courseware they were developing for new laboratory experiments. That helped me to develop a few new CAD laboratory experiments along with a tutorial for using the Mentor Graphics tools in CAD experiments. Subsequently, I introduced the accelerated version of the course in Spring 1991 by modifying some of those laboratory experiments. My colleague Karem Sakallah was instrumental in upgrading and enriching EECS 270 by adopting Xilinx FPGA boards in the late 1990's for the laboratory experiments. As Xilinx Foundation tool used ABEL hardware description language (HDL) that had several limitations compared to popular Verilog HDL that we use in lectures, Karem later replaced them with the Altera FPGA DE-2 board and Quartus II CAD tools. These programmable hardware boards and CAD tools allow students to implement their digital design on a hardware platform, and then test and debug their Verilog design. This is the key part of learning in digital logic design.

In addition to the above-mentioned colleagues, I also taught in team with Kang Shin and Janice Jenkins by sharing our lecture duties. Even though I did not teach in team with Mark Brehob, Valeria Bertacco, Trevor Mudge, Igor Markov, John Myers, Edward Davidson, and David Blaauw, they have also significantly contributed to upgrading the course. I had interacted with all of them now and then about different aspects of the course and shared homework and exams. Matthew Smith manages the laboratory component of EECS 270 quite deftly, making sure that students learn lecture materials of the course through hands-on laboratory experience. In addition to Yalcin Yilmaz, who assisted me in completing few chapters, my research associate, Dr. Mikhail Erementchouk, and two of my doctoral students, Nan Zheng and Soumitra Roy Joy perused the voluminous manuscript and provided me with useful feedback.

Finally, my lectures were also partially influenced by the textbooks we adopted over the course of last so many years. In order to ensure that students who were enrolled in my class could relate to the prescribed textbook, I had assimilated some aspects of my lectures with the styles and contents of these textbooks[1]. I would like to express my sincere thanks to the authors for dedicating themselves to education and training of undergraduate students. The last but not the least, students who undertake this journey to enthusiastically learn computer hardware and apply their knowledge in industry to design and manufacture cutting-edge hardware products are the most important reasons for us, the educators, to continually learn ourselves and assimilate our knowledge in our course offering.

Pinaki Mazumder, Professor
Dept. of Elec. Eng. and Comp. Sci.
University of Michigan, Ann Arbor, USA
April 28, 2019

---

[1]Digital Design (2nd Edition) by Frank Vahid, Wiley, 2010 Digital Design: Principles and Practices (4th Edition) by John F. Wakerly, Prentice Hall Publishing Company Introduction to Digital Logic Design by John P. Hayes, Prentice Hall Logic and Computer Design Fundamentals (3rd Edition) by Morris Mano and Charles Kime, Prentice Hall Publishing Company Fundamentals of Logic Design (5th Edition) by Charles H. Roth, Thompson Brooks/Cole Contemporary Logic Design (1st Edition) by Randy H. Katz, Pearson

List of Figures

List of Tables

List of Abbreviations

BCD	Binary coded decimal
SOP	Sum of products
POS	Product of sums
HDL	Hardware description language
MSI	Medium scale integration
FPGA	Field-programmable gate array
PLD	Programmable logic device
PLA	Programmable logic array
ROM	Read only memory
RTL	Register transfer logic
ASIC	Application specific integrated circuit
EDA	Electronic design automation
APR	Automatic place and route
FSM	Finite state machine
HLSM	High-level state machine
ISA	Instruction set architecture
SRAM	Static random access memory
DRAM	Dynamic random access memory
TTL	Transistor transistor logic
CMOS	Complementary metal oxide semiconductor
CCD	Charge coupled device
APS	Active pixel sensors
IC-CAD	Integrated circuit computer aided design
MSB	Most significant bit
LSB	Least significant bit
MUX	Multiplexer
FF	Flip flop

Introduction

The goal of this lecture is to outline the objectives of the course and provide a spring-board and perspective to the world of digital design.

1.1 Objectives

The goal of this course is to teach design techniques and then apply them in a microprocessor[1] design near to the completion of the course. Digital design exists on many tracks, but the three tracks that will be handled in this course are: theoretical representation through equations and basic gates, implementation options using fabricated hardware, and sequential design techniques.

1.1.1 Equation representation

The chapters under this representation take into account the development of the theoretical framework necessary to understand digital design. First, we begin with number systems to familiarize ourselves with the idea of arithmetic with numbers in different bases. This sets up the idea of arithmetic in binary (base 2). After a few examples, the concepts of number representations and codes are introduced. Numbers can be represented in

---

[1]A microprocessor can be programmed to execute various types of computational tasks without requiring to redesign the hardware since the accompanying digital memories store the application program in the form of binary information units, called bits, and work in conjunction with the microprocessor to perform the intended tasks.

different ways: signed magnitude, 1's complement, 2's complement, etc. We use different number representations to show that some have advantages over others and that choosing one number representation over another in hardware could imply more complex circuitry. Binary coded decimal (BCD) numbers are then used to bridge the gap between the computer world and human understanding; BCD codes are essentially methods devised for hardware to use binary numbers with decimal procedures so that humans can easily trace through arithmetic.

After an extensive introduction into number systems, Boolean algebra is introduced. The Boolean algebra lecture strives to set up the Boolean algebra theory and provide a switching circuit implementation of Boolean algebra. The Boolean expressions (or equations) are the first flavor of logic representation. Boolean expressions are then taken one more level beyond switching circuits to logic gates. The AND, OR, NOT, XOR, and XNOR gates are introduced along with other functions, NAND, NOR, etc. Gate level depiction provides enough introductory information to understand real-world problems that may occur.

The problems are introduced in the timing diagrams lecture which focuses on the role time plays in circuits and why switching time should be an important factor a designer should be aware of when designing circuits. Switching time or circuit delays may cause a theoretically flawless circuit design to exhibit glitches and, therefore, fail in certain instances depending on design specifications. Two types of hazards are mentioned: Static 1 hazards and Static 0 hazards.

The equation representation culminates with the idea of translating behavioral objectives to circuit equations and gate level schematics. The lectures on combinational logic design will introduce the concept of thinking through design problems and providing solutions in the form of equations. These equations or expressions will be of the form of a sum of products or a product of sums. Logic minimization techniques using Karnaugh maps will be introduced to supplement Boolean algebra minimization techniques previously learned. The visual Karnaugh map minimization will only work effectively for less than five or six Boolean variables. Beyond that, the method becomes cumbersome and a tabular method, known as the Quine–McCluskey (Q-M) method, will be discussed toward the end of the course to show that no matter how many input variables exist, the Q-M method can allow us to derive the minimized Boolean equations.

1.1.2 Hardware platform implementation

With the introduction of Boolean algebra, problems with respect to timing inherent in real designs, and logic minimization completed, real-world hardware implementations are considered. This phase of the course deals with implementing Boolean functions using distinct components. It is not as different from the previous section, except now the gates are physical MSI gates. The design in this phase will coincide nicely with the laboratory experiments since MSI chips are used in lab for hardware verification. In addition to using gates to implement Boolean functions, multiplexers and decoders are introduced to further add to the designers' toolbox.

The decoder implementation of Boolean functions is especially interesting because it introduces the basic workings of a read only memory (ROM). This is the first true connection between Boolean functionality and memory. It shows that the canonical sum of products and the product of sums may be obtained by "reading" from memory locations that return either true or false. The ROM cannot be changed and only implements one Boolean function. The extension of the ROM implementation to one that can withstand reprogramming to multiple Boolean functions is accomplished with the programmable logic devices (PLDs). As the name suggests, the PLDs can be programmed to implement different Boolean functions. There are various types of PLDs such as programmable logic arrays (PLA) and the field-programmable gate arrays (FPGA).

The PLA uses programmable AND and OR gates for direct implementation of combinational logic in the canonical sum of products and product of sums forms. It differs from the ROM in that the Boolean logic function can be modified and reprogrammed. As will be seen, the circuits implemented in the canonical form are very expensive in terms of hardware resources. Also as Karnaugh maps explode to over 5 or 6 variables, the ability of the design engineer to effectively minimize a Boolean function by hand is very limited. Therefore, a more compact, intuitive design space called sequential design is used.

FPGAs use a grid of logic gates that can be programmed on field by the end-customer to implement arbitrary logic configurations. They also include storage devices to implement sequential logic. The end-customer can test the initial design of a digital circuit on FPGA, determine possible improvements over the initial design, and make changes on the circuit. The FPGA then can be reprogrammed to test the improved design. FPGAs are crucial in rapid prototyping and reduce the test times and costs.

In very large-scale designs, it becomes impossible to design every single logic block by hand because it would take an infeasibly long time to obtain an initial design. Engineers have created ingenious solutions to overcome this problem by forming conventions that can define the circuits by their desired functionality. These conventions are commonly referred to as hardware description languages (HDLs). HDLs can be used to generate an initial design of a digital circuit by defining the circuit structure and the circuit operations in text. Verilog, VHDL, and SystemC are the most common HDLs used in the industry.

For example, an engineer can first come up with the high level design of a digital system, which he can implement in register transfer level (RTL) using an HDL such as Verilog. RTL is an abstraction level that models the digital circuits by defining the signal flow between registers and the logical operations performed on these signals. The designer can then test and verify the design using software tools. After the initial testing, the designer can synthesize the design specified in Verilog using electronic design automation (EDA) tools. This process converts the RTL level design into a netlist of logic gates. The synthesized design can be further tested for functional correctness and circuit timing using software tools and FPGAs. Finally, the design can be physically laid out for fabrication using automatic place and route tools. The steps mentioned above can be iterative and additional testing may be needed at each step. This design methodology is referred to as application specific integrated circuit (ASIC) design.

It is, therefore, of utmost importance that the digital design engineers should attain the necessary skills to tackle the ever-increasing digital design complexity by learning how to design using an industry standard HDL such as Verilog. However, this does not mean that the manual design skills are no longer necessary. In fact, it is the opposite. Automated tools may not always provide the best performing or optimal design, and high-performance system design requires careful manual design. Therefore, the correct design methodology must be learned systematically. Systematic design methodologies for both combinational and sequential system designs will be provided in the following lectures.

1.1.3 Sequential design

Sequential design may be more compact due to reusing hardware, but it is slower than combinational logic. Sequential design combines combinational logic with memory and will be introduced from simple memory elements (cross-coupled inverters) to more complex and multi-bit memories (flip flops and registers). Timing diagrams with flip flops, counters, and registers are introduced before an in-depth procedure on how to design finite state machines (FSM) using the Mealy machine or the Moore machine.

The FSM design phase will strive to set apart the combinational logic design with the sequential design. Some problems that were tackled using combinational logic will show that they can be implemented with fewer hardware resources in sequential design. As number of states are kept low, the hand design of FSMs are quite useful to help the designer minimize states, but as the problem to be solved becomes more complex and the number of states explodes, then a better method of design is necessary.

The introduction of high-level state machines (HLSMs) is made. HLSMs are considered as problems become more complex, and inputs and outputs are no longer single bits. The HLSM lecture is coupled with Verilog code implementation because it provides a flavor as to how design is actually done in industry. Most digital engineers do not go through the process of minimizing Karnaugh maps or hand minimizing large state tables. The use of HDLs such as Verilog is the key to quicker, robust logic implementations.

The motivation behind the switch to sequential design lies in the fact that as the problem becomes more complex, logic design with Boolean algebra and combinational design becomes very tedious. As FSMs were used to alleviate this tediousness, as the state space grew, FSMs became tedious because the number of flip-flops and their associated combinational gates also grew substantially. Then HLSMs were introduced to handle larger, multibit problems. HLSMs may have fewer states than FSM implementations, but the more abstract we represent and combine states, the farther away from hardware we are moving. In order to implement HLSMs, a higher-level abstraction, called RTL descriptions, is used. There are many ways to design HLSMs, but the RTL method is mostly used.

RTL is important because it can be easily synthesized into logic gates from Verilog description. The design procedure splits up the HLSM into two parts: the datapath and the control. The datapath is primarily designed with combinational logic[2] while the control contains FSMs, synchronized

---

[2]Note that if pipelining is used to improve the system throughput, the combinational logic is partitioned into various stages and each stage then, after computing the output values from the inputs that come from its previous stage, latches its output to a set of flip-flops. Depending on the number of pipelining stages, the system throughput improves considerably. Since the datapath in the case contains an overlapping multiple data from the input stream, this type of overlapping is called temporal parallelism, i.e., time-domain parallelism. If the datapath is

with clock signals that generate handshake and control signals. Handshakes between the datapath and the control blocks are used to synchronize data between both blocks. RTL description enables for better understanding the microprocessor, which is why it is very important. This collection of lectures will be augmented with a few more lectures at the backend.

1.1.4 Datapath components

The later part of the lectures will extend the RTL description into a microprocessor description, utilizing the same concepts as before with the dichotomy between control and datapath. More datapath components, specifically adders and multipliers, will be provided in detail. The adder lecture will build upon full and half adders and then introduce serial and parallel adders. The basis of this lecture is to establish the foundation that there are multiple methods of accomplishing tasks in digital design, and these methods can be classified under serial or parallel methods. After the serial adder is implemented, several parallel architectures are provided, e.g., parallel prefix adders, conditional sum adders, carry select adder, etc. The adder chapter concludes by connecting to an earlier lecture on the subtraction of 2's complement numbers.

With addition and subtraction covered, multiplication and division are introduced. Multipliers are expounded on in detail following the same path as the adder lecture with serial and parallel multiplication. Both encoded schemes for multiplication are discussed for 2's complement multiplication before division is introduced. Although there are serial and parallel dividers, serial dividers are not discussed in these lectures. The reason behind this is that most designers today use serial division. Two main division algorithms will be covered: restoring and nonrestoring algorithms. This division and multiplication lecture essentially covers an essential part of the arithmetic logic unit. The arithmetic logic unit is used in various applications, with the general-purpose processor as the quintessential.

1.1.5 Backend lectures

The backend of these series of lectures holds value in the sense that it combines the knowledge of the previous lectures to explain higher level concepts like the general-purpose processors, Boolean and state

---

replicated multiple times to improve the system throughput, it is called spatial (space-domain) parallelism.

minimization techniques, PLDs, and memory implementation. Besides the minimization techniques lecture, these backend lectures provide potential avenues of specialization for digital designers. The general-purpose processor lecture will build upon the idea of a general-purpose processor with three instructions. Afterwards, the concept of an instruction set architecture (ISA) is introduced. The ISA provides the description necessary to determine what a general-purpose processor can or cannot do. Afterward, an 8-bit general-purpose processor example is described and a sample instruction set is provided.

The general-purpose processor lecture gives room for the minimization lecture, which looks at the Q-M method of minimization of Boolean functions and state minimization techniques when implementing state machines. These minimization tools show systemic ways of performing minimization that allows for writing programs to go through this process. A large part of digital design is being able to automate part of the design process; so knowing algorithms that aid such automation is essential. The Q-M method is compared to the Karnaugh map method from earlier lectures and the state minimization method is compared to the intuitive reasoning of state minimization used in the FSM lectures.

The minimization lectures yield way to the PLD lecture. In this lecture, several fabrics are discussed, specifically, ROMs, PLA devices, complex PLDs, and field PLAs. These fabrics are used to rapidly prototype Boolean functions and are very useful for quickly designing products. After the PLD lecture, a primer on memory is provided. The memory lecture will start with memory hierarchies that may be part of a computing machine or server and give reasons why many different levels of memory exist. Then the static random access memory will be discussed in detail, juxtaposed with the dynamic random access memory (DRAM). A brief mention of other types of memory will then be discussed, focusing on promising future memory that students may encounter during their studies or research. The memory lecture is the final lecture in these series of lectures, so it deals more with devices than other lectures.

1.2 Analog vs. Digital

The design space over the last 40+ years has moved from analog to digital. An analog signal contains many frequencies while a digital signal is a string of 0's and 1's as shown in Figure 1. The *periodic* analog signal can be denoted as a superposition of multiple frequencies whose amplitudes

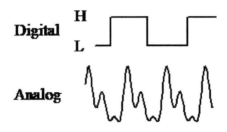

Figure 1.1 Digital and analog signals.

may differ and is often represented by a Fourier series (http://www.fourier-series.com/fourierseries2/Lectures/ Intro_to_FS_lecture/index.html).

Digital signals normally take two values that are denoted as "Logic 0" and "Logic 1." These can be different for different technologies, for example, transistor–transistor logic values for Logic 0 at the output of logic gates are voltages <0.4 V and Logic 1 is between 3.6 and 5 V. However, modern complementary metal oxide semiconductor (CMOS) technology uses a transistor and wire shrinking mechanism to scale down the supply voltage progressively with the goal to minimize power consumption. This scaling mechanism has allowed CMOS technology to grow the number of transistors in a chip to several billion transistors within 1 cm^2 of the silicon area and the power supply voltage is typically between 700 mV and 1 V. Digital signals are grouped into two levels, making them much simpler to deal with than analog signals. Analog signals can take on values from 0 V to whatever the power supply may provide. Analog signals for over-the-counter components may require voltages between 15 and –15 V, i.e., a 30-V differential. Over-the-counter digital components usually do not require such a large differential. Power consumption for analog circuits is, therefore, usually larger than digital circuits.

In addition to power consumption, digital design has an advantage over analog design in terms of density. Analog design usually involves very careful design and methods that increase chip area. Since digital design is only concerned with two logic levels, as long as these levels can be deciphered, we do not care if the signal level is 0.1 or 0.09 V. Some simplification can be made in digital, thereby allowing for simpler designs than their analog counterpart. Simpler designs also mean better automation; so digital designs can be generated by a computer program, while analog designs cannot. Also by reducing the precision of a system to two levels, digital designs can be operated much more quickly than analog design; hence, they are usually much faster. It must be noted that faster circuits entail faster processing; so analog processing that can be done digitally has quickly shifted to digital, hence the ubiquitous scene of digital systems today. A few examples of applications that were once analog and shifted to digital are provided below:

Still Photography: NASA has been very instrumental in the development of the digital photography. They wanted an alternative to the use of vacuum tubes on satellites for astronomy. The answer they sought was in the development of the charge-coupled device (CCD) in Bell Laboratories. This invention allowed the revolutionary explosion of digital photography (Figure 1.2). Today, CCD is on its way out and active pixel sensors with better resolution, smaller size, and cheaper manufacturing costs are used in digital cameras. Digital photography is definitely more prevalent than its analog counterpart today because they are easy to transmit over long distances, easier to store, and do not require hours for film exposure.

Figure 1.2   An Olympus digital camera and a picture from it stored as compressed digital bit strings.

Sound and Video Recording: Just like photography, video recording has changed significantly over the years. Today, Blu-Ray and digital video discs (DVDs) are used to store video digitally. With the invention of the DVD player and disks, analog video tapes such as video home system (VHS) market were wiped out (Figure 1.3). The same pattern was observed after CDs were invented. Although tape players stuck around much longer, their longevity coincided with the lack of accessible CD burners. When CD burners were more available to the consumer, the audio tape industry saw its last throes.

Watches: Analog and digital watches have managed to coexist and one has not completely taken over. Although analog watches were first invented,

Figure 1.3   High-density DVD and analog VHS tape.

digital watches with their added functionality have been unable to make them invalid. The reasons may be superficial to the point of marketing, preference, and perception. Analog watches are usually viewed as more suave than the digital watch, which is more casual and sporty (Figure 1.4).

Figure 1.4   A Rolex analog watch is priced at $10,000 while its digital counterpart may cost only $150.

Television: TV sets and TV signal transmission have moved to the digital domain. In the U.S., analog channels are no longer available. In Ann Arbor, you may be able to get a couple of analog Canadian broadcast TV with an antenna, but no more analog broadcasting is made in USA. Hence, television production today is far friendlier to digital signals. For example, with an antenna, the TV is able to show what shows will be on next and provide descriptions of shows. Display technology today has phased out the CRT monitors, and high definition televisions use active matrix LCDs, plasma, and organic LEDs. Whereas the black-to-white contrast ratio that delineates the picture quality attainable in CRT displays is in upper 100s, the contrast ratio with LCD TV is about 50,000, with new generation LED back-lit LCD TV is about 1,000,000 and it is above 2,000,000 with organic LED TV. The current dominant digital TV manufacturers are Samsung Electronics, Panasonic, Toshiba, Philips, LG Electronics, ProScan, Kogan, Sony, and Vizio. They have pushed the TV technology beyond high-definition (HD), and currently they have introduced more advanced designs such as ultra-high-definition (UHD), three-dimensional (3D), and web television.

Telephony: Cordless telephones became digital because digital signals are easier to compress, transmit, encode, and decode compared to analog signals (Figure 1.5). Analog cordless phones would be choppy over a long distance while talking; so, virtually, all cordless phones today are digital. Cellphones became an extension of the cordless phone principle and so are all devices that work with digital signals. The companies who still dabble with analog telephony are in niche markets and find profit in the fact that they do not need to sell very many analog telephones because the production cost for each telephone is cheaper than the digital phone.

Figure 1.5 From the Gower model (1912) to modern cell phones: the phone industry went digital.

1.3 Digital Design

The objectives of the course have been laid out and many examples have shown how digital design for several important applications has overtaken analog design. The truth of the matter is digital design lies in a large space and requires the amalgamation of different skills in order to take an idea or specification and turn it into a product. This section will provide three broad levels (behavioral, structural, and geometric) of digital design and then attempt to provide a picture of what digital designers actually do.

1.3.1 Levels of digital design

Behavioral

Behavioral level lies in using English words, equations, or programming languages to describe circuit actions and activities. Included in this level are FSM flow descriptions, Boolean equations, RTL design, and using HDLs such as VHDL or Verilog to describe circuit block behavior. An example of a Verilog behavioral description of an arithmetic adder unit that provides the digital sum of two binary bits along with an input carry bit is shown in Figure 1.6. Inputs and outputs are defined and the code essentially uses an equation to relate the inputs to the output.

```
module add2bit (A, B, Cin, S);
      input [1:0] A, B;
      input Cin;
      output reg [1:0] S;

      always @ (A or B or Cin)
         begin
            S = A+B+Cin;
         end
endmodule
```

Figure 1.6   Example of a behavioral description of a 2-bit adder in Verilog.

Structural

Structural level lies in creating a schematic representation out of behavioral descriptions. There are different flavors of structural implementation, e.g., gate level schematics, transistor level schematics, and leaf cell representation of circuits. The structural level uses actual connections and wires, and the signals that flow are voltages and currents. An example of a gate level structural representation of a Boolean function is shown in Figure 1.7. In this example, OR gates are used to combine signals to provide three outputs which are combined with an AND gate to provide the output Z. The structural representation is essentially expressing the behavioral description that $Z = (A + B)(B + C)(A + C)$.

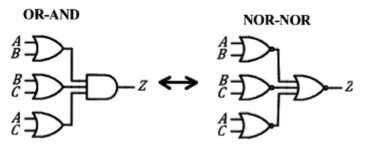

Figure 1.7   Example of a structural description of a circuit implementing $Z = (A + B)(B + C)(A + C)$.

Geometric
Geometric representation deals with converting the structural level representation into physical layers (Figure 1.8). The chip floorplan, cell and module placements, and mask generation are involved at this level. This is the closest level to the actual manufactured product because it provides the information on how the chip should be fabricated. An example of a geometric representation is given in Figure 8. Without proper training, it is difficult to understand geometric representations. The rectangles in different colors correspond to different levels but do not worry about the details; just know that the CMOS technology is used to layout transistors, and this is an example of what someone doing layout for a chip will work with.

1.3.2 What do digital designers do?

With the vast expanse of the three levels already presented, it is difficult to know what classifies a person as a computer engineer. The crux of the matter lies with the uncertainty involved in what computer engineers or, for that matter, digital designers actually do. We have laid out the objectives of the

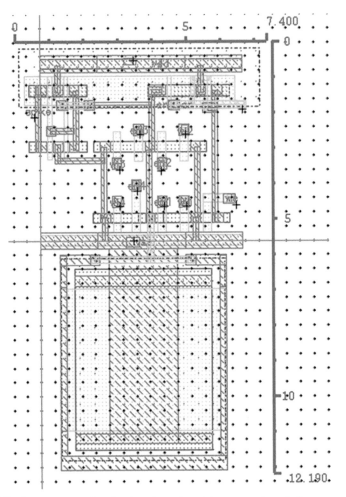

Figure 1.8 Layout example showing how transistors are viewed from a geometric perspective.

course, which covers the core subjects all computer engineers should know. But the question remains in how this translates into actual jobs and tasks for engineers after college. Digital design is bundled into different flavors and so are their designers. From a broad perspective, digital designers do two things: the first is they design digital systems, and the second is they develop tools for digital design. We will first discuss the first and then the second.

It seems puerile and dissatisfying to ask, "What do digital designers do?" and then provide the answer, "Digital designers design digital systems." So we will clarify what is meant by systems. For example, some digital designers are computer architects and specialize in designing processors. Some are ASIC designers and specialize in quickly producing custom digital designs. Some are infused in hardware and software and enjoy embedded system design. Others are interested in digital electronic design and are, therefore, called digital electronic circuit designers. These may all sound the same, but the specialties have different focuses in complexity, hence the distinctions.

General-purpose microprocessors such as Intel Pentium and AMD Turion have grown in speed, and as they are becoming faster, their complexity has exploded. The microprocessor is plagued with complex memory and core interactions, and with the move to multi-core processors with shared cache, processor design does not seem to be getting any easier. Digital designers love microprocessors because they entail a huge design problem on all levels of digital design, including compiler and electrical and manufacturing problems. Microprocessors are frequently aggregated to build large clusters of processors that can perform high-performance or supercomputing at a trillion operations per second rate generally known as Petascale computing. The next generation Exaflops Data Center supercomputer will require at least 223,872 microprocessor chips, 3.58 million DRAM chips, and about 300,000 hard disks to compute at nearly one Exaflop (10^{18}) rate, while a departmental parallel computer will perform just over a Petaflop (10^{15}) rate requiring about 384 microprocessors, 6144 DRAM chips, and 512 hard disks.

ASICs tend to be a much smaller design than the microprocessor and usually target a specific application. Most of the designs are performed with a small team of individuals who tend to work quickly to get a design out to market. They depend extensively on the automation of CAD tools. Digital designers who tend to go for ASICs admire the need to solve different interesting problems from a given specification or targeted behavior. ASIC design may be a smaller design than the microprocessor, but its complexity lies in finding the best solution for the given lead-time; hence, it involves a big problem in design strategy.

Embedded system design tends to work differently from the other two. Embedded systems usually are a combination of a processor, memory, custom logic, and software. Depending on the complexity of the embedded system, most companies do not produce all four components in-house. Embedded system design usually starts with the processor, then custom logic and/or more memory besides that in the processor is added, and then the embedded software is added on. Embedded system design, just like ASIC design, is made for a specific purpose. These include multimedia chips, cellular telephones, car systems, gaming systems, etc. Designers involved on the software side of an embedded system design need to know the hardware in order to know what the hardware is capable of. Software design for embedded systems is different from that for the general-purpose computer.

The second thing digital designers do is to develop tools for digital system design. This is a very big area that continues to grow. EDA, also known as integrated circuit computer aided design (IC-CAD), software tools have enabled quickly designing circuits to market and reduced the time and money costs of a circuit design. EDA tools exist in all levels of digital design: there are IC-CAD tools for generating chip layout, IC-CAD tool improvements that allow generating circuit schematics from behavioral HDLs such as Verilog, VHDL and System-C, and IC-CAD verification and simulation tools that check for potential problems that may arise from a certain design. In order to build these EDA or IC-CAD tools, the tool designer must understand digital hardware design.

A design space that encompasses these two activities of digital designers is *testing and debugging.* The testing and developing methodology for testing is such a huge market. As designs have exploded in complexity, old methods of testing and brute force testing would take too long, so testing is always an area where design engineers are needed. With the development of IC-CAD tools and reusing of already created hardware intellectual property, the need for custom designers is lower than that for testing engineers. As you learn design principles, keep in mind that within computer engineering, specialization occurs based on interest, and there are ways to set yourself apart from others. The field is just that big!

Numeral Systems and BCD Codes

This chapter formally introduces the numeral system commonly used today (positional numeral system) and compares it with the unary numeral system. Addition and subtraction procedures are reviewed for the positional numeral system, and hexadecimal, binary, and octal arithmetic are discussed. Afterward, the binary representation of negative numbers is discussed, specifically the signed magnitude system, 1's complement representation, and 2's complement representation. Addition and subtraction with these representations are discussed in detail. Finally, binary-coded decimals are introduced as a framework for the human understanding of binary 1s and 0s.

Terms introduced in this chapter: unary numeral system, positional numeral system, decimal, base, radix, decimal point, radix point, binary, hexadecimal, octal, bit, binary point, most significant bit, least significant bit, nibble, signed magnitude, diminished radix complement, 1's complement, radix complement, 2's complement, self-complementing code, reflective unit distance code

Competency Objectives: At the end of this chapter, you will be able to:

1. Distinguish between unary and positional numeral systems.
2. Convert a number from an arbitrary base p to another arbitrary base q.
3. Add and subtract numbers in an arbitrary base.

4. Perform addition and subtraction using signed magnitude, 1's complement, and 2's complement representations of negative numbers in binary.
5. Learn various binary-coded decimal representations.

2.1 Introduction

2.1.1 Unary numeral systems

Over generations, several numeral systems came into being before the positional system became ubiquitous. The simplest numeral system is the *unary numeral system*; this number system works by repeating characters for larger numbers. For example, the tally marks system still utilized today in many game settings is shown in Figure 2.1a. Each character is represented by a forward slash, "/," and the number of slashes indicates the value of the number represented. A more sophisticated unary numeral system, even though not commonly used today, is the Roman numeral system, as shown in Figure 2.1b.

Unary systems have been used throughout history, with some of its variants including the Egyptian and Chinese numeral systems. The problem with unary numeral systems is that they become very cumbersome when they deal with large numbers. For example, try computing the following arithmetic without converting to the decimal system you are used to: (XLII × MC) – (XXI × MMCC). Compare that to the following equivalent calculation: (42 × 1100) – (21 × 2200). The ease of working with the positional numeral system has contributed to its widespread use.

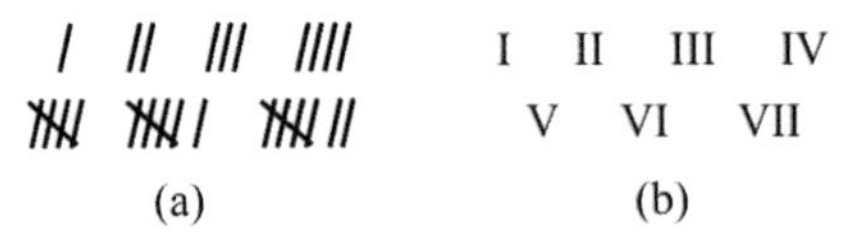

/	//	///	////	I	II	III	IV
⋕	⋕/	⋕//		V	VI	VII	
(a)				(b)			

Figure 2.1   (a) Tally mark counting to 7 and (b) Roman numeral counting to 7.

2.1.2 The positional numeral system, or the place-value system

The *positional numeral system*[1] is the common number system used. For example, the number 1385.3 is interpreted as $(1 \times 1000) + (3 \times 100) + (8 \times 10) + (5 \times 1) + (3 \times 0.1)$. Each digit in the number 1385.3 is multiplied by a different place value and added together in order to obtain the value of the number. The given number 1385.3 is in *decimal* format because every place value is a power of 10. Hence, 1385.3 is therefore interpreted as

$$1385.3 = (1 \times 10^3) + (3 \times 10^2) + (8 \times 10^1) + (5 \times 10^0) + (3 \times 10^{-1})$$

In this instance, 10 is called the *base* (or *radix*) of the numeral system. In general, a number can be written in any base greater than or equal to 2. Note that, after the place value of 10^0 power, a *decimal point* is placed to signify the start of the negative exponents of the base. In a general base, this point is called a *radix point*.

In conclusion, the place-value system allows the representation of numbers in any base through a weighted sum of the digits of the number with exponents of the base. A radix point is used to separate whole parts of the number from the fractional parts. The most commonly used bases are *decimal* (base 10), *binary* (base 2), *hexadecimal* (base 16), and *octal* (base 8).

2.1.2.1 Binary numbers

Binary numbers are numbers written in base 2. They can be represented only with the digits 0 and 1. Binary digits are called *bits*, and the radix point in a binary system is called a *binary point*. When writing numbers in a base other than base-10, it is customary to write the base as a subscript. For example, the number 1101_2 clearly shows that this number should not be mistaken for "one thousand one hundred one." Instead its value is $(1 \times 2^3) + (1 \times 2^2) + (0 \times 2^1) + (1 \times 2^0) = 13$. So $13_{10} = 1101_2$. The only time the base is not written is when the base is clear from the context.s. The leftmost digit of a binary number is called the *most significant bit* (*MSB*), and the rightmost digit of a binary number is called the *least significant bit* (*LSB*).

---

[1]The Babylonian sexagesimal (base-60) system, which was supposedly inherited from Sumerian civilization (c. 4500c. 1900 BC) is credited as being the first positional numeral system (PNS). The Babylonian system (c. 2000 BC) used only two distinct symbols similar to I and < to count unit and ten, respectively. Therefore, ⋘ II would represent 32 in the Babylonian numeral system and those two symbols were aggregated to denote 59 numerals. The base-60 and base-20 number systems were popular with ancient civilizations like Egyptian, Chinese, and Hellenistic Greek because 60 is a common number in geometry (180 degree and 360 degree), time (60 minutes and 60 seconds), and so on.

Binary numbers are important because digital systems use the binary system much more efficiently than other number systems. Conventional digital computers use two different voltage levels: supply voltage (V_{DD}) and a ground voltage (V_{DD}) to represent 1 or High (H), and 0 or Low (L) − two different symbols needed for binary calculation.

2.1.2.2 Hexadecimal numbers

Hexadecimal numbers are numbers expressed in radix 16. These numbers need 16 different symbols or digits, but the decimal digits only have 10 symbols, that is, 0 through 9. To make up for the other 6 symbols, the letters A through F are used as digits to signify values 10 through 15. For example, $5A9_{16}$ is equivalent to the expression $(5 \times 16^2) + (10 \times 16^1) + (9 \times 16^0) = 1449$.

Hexadecimal (or hex) representation is important in digital computing because it provides a shorthand notation for programmers and designers to write binary numbers. Each hexadecimal digit in a number conveniently represents a 4-bit binary number called a *nibble*. To convert from hexadecimal to binary, simply convert each hexadecimal digit to a 4-bit binary number. In our example, $5A9_{16}$ in binary is as follows:

$$(5\ A\ 9)_{16}$$

$$(0101\ 1010\ 1001)_2 = (10110101001)_2$$

where 0101 is 5 in binary, 1010 is A in binary, and 1001 is 9 in binary. Combine these and you get your hexadecimal to binary conversion. To convert from binary to hexadecimal, do the opposite process: group the binary digits into groups of four (starting from the LSB) and convert each group to a hexadecimal digit. Add leading zeros, if necessary. For example, $11111000001101010_2 = 0011\ 1111\ 0000\ 0110\ 1010_2 = 3F06A_{16}$. Note that if a programmer who is writing code for a binary digital computer to reference a memory location whose 16-bit binary address is 1001 0011 1110 1010, it is cumbersome to write the string of 0s and 1s. It is much easier to write in hexadecimal representation, which is $93DA_{16}$.

2.1.2.3 Octal numbers

Octal numbers are numbers in base 8. These numbers are represented with 8 digits, that is, 0 through 7. An example of an octal number is 732_8, which is equal to $(7 \times 8^2) + (3 \times 8^1) + (2 \times 8^0) = 474$.

In the octal number system the radix, $8 = 2^3$, is a power of the binary radix, 2. This relation leads to the advantage of the ease of conversion between octal and binary numbers. To convert from octal to binary, simply convert each octal digit to a 3-bit binary number (recall that, conversion from hexadecimal to binary numbers requires each hex digit to be converted to 4-bit binary number). In our example, 732_8 in binary is as follows:

$$732_8 = 1110110102 = 1110110102$$

where 111 is 7 in binary, 011 is 3 in binary, and 010 is 2 in binary. To convert from binary to octal, the reverse process is followed: group the binary digits into sets of 3, starting from the LSB, and convert each set to a digit. Add leading zeros, if necessary. For example, $111111000001101010_2 = 111\ 111\ 000\ 001\ 101\ 010_2 = 770152_8$.

2.1.2.4 Converting from decimal to different bases

The decimal number system, because of its prevalence, is easier for human interpretation. While 732_8 and 474_{10} are equivalent, the latter number is more familiar to us. When we evaluate the weighted sum of the digits of a number written in an arbitrary base, we essentially convert the given number to a decimal base.

The reverse process of converting from decimal to any base follows an iterative procedure that involves divisions and finding remainders. The procedure is to take the decimal number and divide it by the base of interest. For instance, if converting from decimal to binary, we take the decimal number and divide by 2. In division, the number being divided is the dividend, the number that divides is the divisor, and the result is the quotient. The decimal number in this example is the dividend, the number 2 is the divisor, and the quotient will be of the form of an integer quotient with a remainder. The remainder will yield the LSB of the base-2 number. The integer quotient will be used as a dividend for the next iteration. Continue the division process and collect remainders until you arrive at "0 remainder X," where X is an integer. This process is best explained using an example, so we show the conversion of 203 to binary, hexadecimal, and octal in Figure 2.2.

The conversion procedure also applies to numbers in radix point. For example, Figure 2.3 shows the conversion of 42.625 from decimal to binary, hexadecimal, and octal. The whole part is treated as in the previous example, but the fractional part is treated differently. The fractional part proceeds through a series of multiplications: After each multiplication step, the whole part of the answer (left of the radix point) is kept as a digit, while the

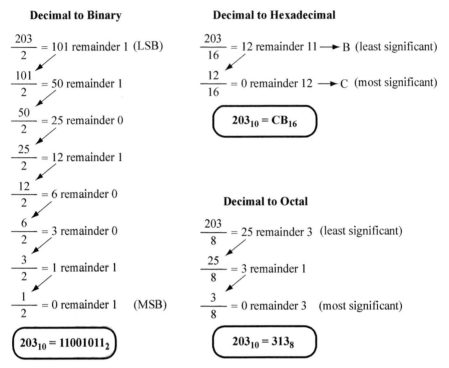

Figure 2.2 203 is converted to binary, hexadecimal, and octal to illustrate the conversion procedure.

remaining fractional part undergoes further multiplication. The process is repeated until the fractional part is 0. The conversion process is exemplified below.

2.1.2.5 Converting from one arbitrary base (p) to another arbitrary base (q)

Once we know the conversion mechanism between a number in the decimal base and that in some other base, we can leverage the technique to convert between any two arbitrary bases by using base-10 as an intermediary.

To convert a number from base p to base q, follow the two-step guideline shown in Figure 2.4. First, convert the base p number to base 10 and then to base q.

Note that, as in the decimal system, there is the possibility of repeating digits in the fractional representation. In the example, while converting the fractional part of 39.68 to base 7, a repeating pattern is observed.

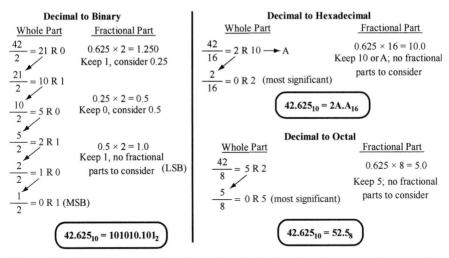

Figure 2.3 42.625 is converted to binary, hexadecimal, and octal to show the conversion procedure when a radix point is involved.

The concept of number systems is important for the design of more efficient computers and high-density data storage. Multivalued computers are expected to reduce the wiring length in a microprocessor and also make the logic functions more compact, thereby improving performance both in terms of speed of computation and energy dissipation. Nonvolatile memories such as Flash increase data storage density by deploying multi-level cells (MLC) capable of storing 3-bits/cell.

Peta-scale cloud computers, having speed above 10^{15} Floating Point Operations Per Second (FLOPS), presently consume over 50 Mega Watts (MW) of power, and this may escalate to nearly a Giga Watt (GW), if datacenters and supercomputers run at 10^{18} FLOPS in future Exa-scale systems. In order to reduce power consumption, future supercomputers may use multi-level logic (say, of base p). Similarly, a laptop may also use multi-level logic (say, of base q) to save battery power, thereby creating the need for data portability in mixed computing systems.

2.2 Addition and Subtraction in the Positional Numeral System

The addition and subtraction algorithm in an arbitrary base can be similar to that we follow in the decimal base. An example is shown in Figure 2.5, where 136 is added to 191 in different bases.

The concept of *carry* is preserved from one base to another. When a system exceeds its largest digit, a carry is generated. In the decimal case $9 + 3 = 12$, in binary $1 + 1 = 10$, in hexadecimal $8 + F = 17$, and in octal $1 + 7 = 10$. Another example of addition is shown in Figure 2.6 for more practice. Trace through these examples to convince yourself that the arithmetic is correct.

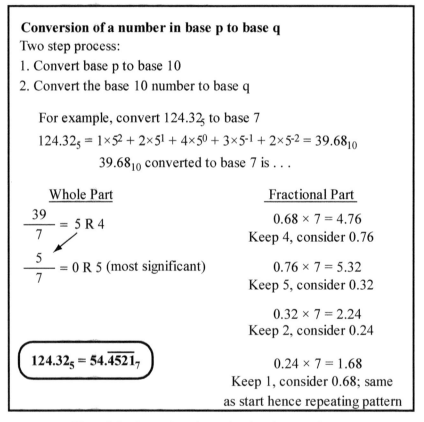

Conversion of a number in base p to base q

Two step process:

1. Convert base p to base 10
2. Convert the base 10 number to base q

For example, convert 124.32_5 to base 7

$$124.32_5 = 1\times5^2 + 2\times5^1 + 4\times5^0 + 3\times5^{-1} + 2\times5^{-2} = 39.68_{10}$$

39.68_{10} converted to base 7 is . . .

Whole Part

$\dfrac{39}{7} = 5\,R\,4$

$\dfrac{5}{7} = 0\,R\,5$ (most significant)

$124.32_5 = 54.\overline{4521}_7$

Fractional Part

$0.68 \times 7 = 4.76$
Keep 4, consider 0.76

$0.76 \times 7 = 5.32$
Keep 5, consider 0.32

$0.32 \times 7 = 2.24$
Keep 2, consider 0.24

$0.24 \times 7 = 1.68$
Keep 1, consider 0.68; same
as start hence repeating pattern

Figure 2.4   Conversion of a number from base p to base q.

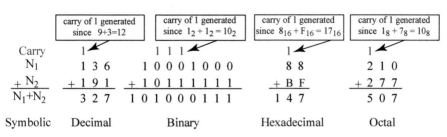

	carry of 1 generated since $9+3=12$	carry of 1 generated since $1_2+1_2=10_2$	carry of 1 generated since $8_{16}+F_{16}=17_{16}$	carry of 1 generated since $1_8+7_8=10_8$
Carry	1	1 1 1	1	1
N_1	1 3 6	1 0 0 0 1 0 0 0	8 8	2 1 0
$+ N_2$	$+$ 1 9 1	$+$ 1 0 1 1 1 1 1 1	$+$ B F	$+$ 2 7 7
N_1+N_2	3 2 7	1 0 1 0 0 0 1 1 1	1 4 7	5 0 7
Symbolic	Decimal	Binary	Hexadecimal	Octal

Figure 2.5   136 is added to 191 in decimal, binary, hexadecimal, and octal. Addition involves digit-by-digit addition starting from the least significant digit.

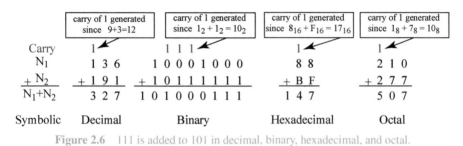

	carry of 1 generated since $9+3=12$	carry of 1 generated since $1_2+1_2=10_2$	carry of 1 generated since $8_{16}+F_{16}=17_{16}$	carry of 1 generated since $1_8+7_8=10_8$
Carry	1	1 1 1	1	1
N_1	1 3 6	1 0 0 0 1 0 0 0	8 8	2 1 0
$+ N_2$	$+$ 1 9 1	$+$ 1 0 1 1 1 1 1 1	$+$ B F	$+$ 2 7 7
N_1+N_2	3 2 7	1 0 1 0 0 0 1 1 1	1 4 7	5 0 7
Symbolic	Decimal	Binary	Hexadecimal	Octal

Figure 2.6   111 is added to 101 in decimal, binary, hexadecimal, and octal.

In subtraction, the concept of borrowing is shown in Figure 2.7. 75 is subtracted from 182 in decimal, binary, hexadecimal, and octal.

These subtractions show the adjustment of borrow terms by striking out the minuend (N_1) multiple times. It can be represented more clearly as shown in Figure 2.7.

Figure 2.7, where 75 is subtracted from 182. Usually, when a larger digit cannot be subtracted from a smaller one, 1 is borrowed from the next place value. In decimal, the value of the borrowed 1 is actually 10; in binary, the value is 2; in hexadecimal, the value is 16; and in octal, the value is 8.

These subtractions show the adjustment of borrow terms by striking out the minuend (N_1) multiple times. It can be represented more clearly as shown in Figure 2.7.

N_1	1 $\cancel{8}^7$ 2	$\cancel{1}^1 0 \; 1^0 \cancel{1}^1 0 \; \cancel{0}^{\,0}\cancel{0}^{10} \cancel{1}^1 0$	$^A\cancel{B}^{16}$	2 6 6
$- N_2$	$-$ 7 5	$-$ 1 0 0 1 0 1 1	$-$ 4 B	$-$ 1 1 3
N_1-N_2	1 0 7	1 1 0 1 0 1 1	6 B	1 5 3

Figure 2.7   75 is subtracted from 182 in decimal, binary, hexadecimal, and octal.

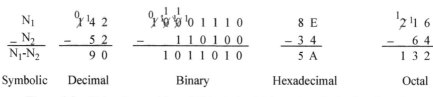

N_1	$0 \atop \not{1}$ $\overset{1}{\not{4}}$ 2	$0 \atop \not{1}$ $\overset{1}{\not{0}}$ $\overset{1}{\not{0}}$ $\overset{1}{\not{0}}$ 0 1 1 1 0	8 E	$1 \atop \not{2}$ $\overset{1}{\not{1}}$ 6
$- N_2$	$-$ 5 2	$-$ 1 1 0 1 0 0	$-$ 3 4	$-$ 6 4
N_1-N_2	9 0	1 0 1 1 0 1 0	5 A	1 3 2

Symbolic	Decimal	Binary	Hexadecimal	Octal

Figure 2.8   52 is subtracted from 142 in decimal, binary, hexadecimal, and octal.

We should be careful about the value of the borrowed 1 when we subtract at a given base. This is especially important in binary when multiple borrowings occur, as in Figure 2.8, where 52 is subtracted from 142. In decimal, when 1 is borrowed from 10, a 9 is left in its place, as when doing $10 - 1$. In binary, when 1 is borrowed from 10, a 1 is left in its place because $10 - 1 = 1$. Trace through the binary subtraction in Figure 2.8 to make sure the process is well understood.

We have added and subtracted only positive numbers so far and have not mentioned negative numbers. There are multiple ways of dealing with negative numbers, as will be shown in the next section. Also to note, division and multiplication in the positional number system are as taught in elementary school and will not be addressed in this chapter. Binary multiplication and division require more advanced knowledge in combinational circuits and finite-state machines; especially, digital hardware often uses sequential or serial multiplication and division since they lower hardware cost at the expense of sacrificing speed of operation. On the other hand, in high-performance arithmetic circuits, a parallel multiplier and divider are used at the cost of the proliferation of the number of gates with an increase in the number of bits for multipliers and divisors.

2.3 Negative Numbers in Binary

There are many negative number representations in binary, but only three will be discussed: signed magnitude, 1's complement, and 2's complement.

2.3.1 Signed magnitude

The *signed magnitude* negative number representation is the most recognizable in everyday life because it is the one we use daily. A sign denotes whether a number is positive or negative. If no sign exists, then the number is assumed to be positive. For example, looking at the numbers +7,

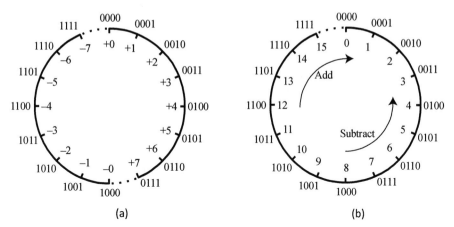

Figure 2.9 Graphically comparing 4-bit (a) signed magnitude numbers with (b) unsigned binary numbers.

−7, and 6, we can tell +7 and 6 are positive numbers and −7 is a negative number. The signs "+" and "−" are used to differentiate a positive number from a negative one.

In the signed magnitude representation of a binary number, the MSB is used as the sign bit. Instead of "+" and "−," 0 and 1 are used to represent positive and negative, respectively. For example, the number +5 can be represented as 0101, and −5 can be represented as 1101. During the conversion process, the MSB is translated to either "+" or "−," and the other digits are used in the place-value conversion.

Figure 2.9a provides a 4-bit signed magnitude example showcasing all the numbers that can be represented using this scheme. Alongside, Figure 2.9b provides a contrast to show unsigned 4-bit binary numbers. The signed magnitude representation for 4 bits can code the numbers from −7 to +7. Hence, in signed magnitude, n bits can code decimal numbers from $-(2^{n-1} - 1)$ to $+(2^{n-1} - 1)$. Also notice that there are two representations of 0, that is, +0 and −0 or 0000 and 1000.

The signed magnitude is usually not used for arithmetic in digital logic, because it requires complex circuitry. Although we are good at signed magnitude addition of negative numbers, there are too many steps involved. For example, to add two numbers X and Y, first, we check the sign for X and then for Y. If the signs of both numbers are the same, then we add both numbers and preserve the sign. If the numbers have different signs, then we must check which number has a larger magnitude. Afterward, we subtract the smaller number from the larger number. The result will take on the sign of the larger-magnitude number. This additional process involves too many checks and conditions by a binary block known as a *signed magnitude comparator*, which will be discussed in a later chapter. Therefore, instead of signed magnitude, other negative representations are more commonly used in digital arithmetic.

2.3.2 The 1's complement

The *1's complement* negative number representation is classified under a numbering scheme called the *diminished radix complement*. Given an n-digit number X_1 in base r, we want to find another n-digit number X_2 that when added to X_1 will yield $r^n - 1$. Therefore,

$$X_1 + X_2 = r^n - 1$$

where X_1 is the starting value and X_2 is its diminished radix complement. If X_1 were a positive number, then X_2 is considered its negative counterpart, and vice versa. This complement representation works for all bases, but when the base $r = 2$, then the representation is called 1's complement.

Given that the number X_1 is a positive number and we want to find X_2, then X_1 must be subtracted from $r^n - 1$. In binary, this is a simple process because the number $r^n - 1$ is an n-bit number with all digits equal to 1. For example, in a 4-bit representation, $r^n - 1$ is $10000 - 1 = 1111$. Subtracting a binary number from a series of 1's equates to complementing the bits of the original number.

The 1's complement representation for 4 bits can code the numbers from -7 to $+7$, as shown in Figure 2.10. Similar to the signed magnitude, n bits can code $-(2^{n-1} - 1)$ to $+(2^{n-1} - 1)$. Two representations of 0 exist in the 1's complement.

The 1's complement is convenient because of the symmetry between numbers. To obtain a negative number, complement each bit, and you are done. Starting from -7 and adding 1 corresponds nicely in binary addition along the wheel. The only hiccup is $1111 + 1$ will yield 0000. The 1's complement adder will, therefore, need a method to avoid the problem whereby 1 added to -0 is $+0$. Even with this added complexity, this adder is still much simpler than the signed magnitude adder.

Although the 1's complement scheme seems far from the place-value system, the weighted-sum approach can be used to convert a 1's complement number to a decimal number. The change in approach is that the weight of

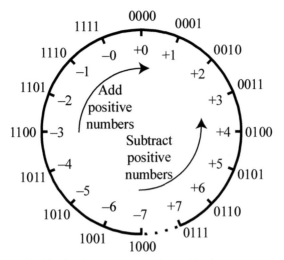

Figure 2.10   Graphically showing all numbers in a 4-bit 1's complement representation.

the MSB is no longer 2^{n-1} but $-(2^{n-1} - 1)$, as shown in the equation:

$$\text{Decimal value} = -\left(2^{n-1} - 1\right) \times a_{n-1} + 2^{n-2} \times a_{n-2} + \cdots + 2^0 a_0$$

An example is used to illustrate the use of the 1's complement. Normally the 1's complement works with a set number of bits. In our example, we choose an 8-bit representation for decimal numbers 59 and 67.

Decimal number	59	67
Binary representation	111011_2	1000011_2
8-bit 1's complement	00111011_2	01000011_2
Negation (complement the bits)	11000100_2	10111100_2
Decimal verification of **negation**	$-(2^{8-1} - 1) + 2^{8-2} + 2^{8-6}$ $= -(2^7 - 1) + 2^6 + 2^2 = -59$	$-(2^{8-1} - 1) + 2^{8-3} + 2^{8-4}$ $+ 2^{8-5} + 2^{8-6} = -(2^7 - 1)$ $+ 2^5 + 2^4 + 2^3 + 2^2 = -67$

Here is a summary of the properties of the 1's complement:

- Complement the bits to change from a negative number to a positive number, and vice versa. It is a symmetric number system.
- As in the signed magnitude system, if 0 is the MSB, then the number is positive, but if 1 is the MSB, then the number is negative.
- There are two representations of 0, +0, and –0. Negating –0 returns +0 and vice versa.

			1	
2	0 0 1 0	5	0 1 0 1	
+ 5	+ 0 1 0 1	+−6	+ 1 0 0 1	
7	0 1 1 1	−1	1 1 1 0	

Figure 2.11   4-bit 1's complement addition showing $2 + 5 = 7$ and $5 + (-6) = -1$.

- An n-bit 1's complement number can represent $-(2^{n-1} - 1)$ to $+(2^{n-1} - 1)$ in decimal.
- Positive numbers have the same coding as signed magnitude positive numbers.

2.3.2.1 Addition in the 1's complement

The 1's complement addition is straightforward and follows binary addition. With the graphical wheel in Figure 2.10, we are just moving either to the left or to the right, depending on the signs of the numbers being added. Two examples of addition are shown in Figure 2.11.

Overflow: The algorithm of addition in 1's complement involves some special cases that require additional maneuvering: for instance, what to do when there is a carry-out out of the MSB position and what to do when an overflow occurs. The 1's complement works for a set number of bits, and sometimes addition will result in a carry-out from the MSB. When this occurs, add the carry-out to the result[2]. This process is shown in Figure 2.12 for a 4-bit example.

For an overflow to occur, referring back to the 4-bit graphical wheel in Figure 2.10, the addition operation must cross the dotted region between −7 and +7 in either direction. An overflow can be easily detected, though. When two numbers of the same sign are added together and the result produces a different sign, then an overflow has occurred. For example, in Figure 2.13, when 5 and 6 are added, the expected result is 11, but instead, the addition yields −4. Also, −5 and −4 are added, but the result is +6, so we know an overflow has occurred.

---

[2]Note that addition in 1's complement may increase the addition time by a factor of 2 in the worst case. Therefore, 1's complement addition and subtraction are not usually implemented in digital hardware. As discussed later, 2's complement overcomes this problem and is generally used in building the arithmetic unit of a digital computer.

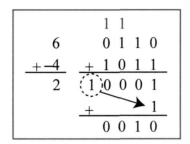

Figure 2.12   Example showing what to do when an addition yields a carry-out from the MSB position.

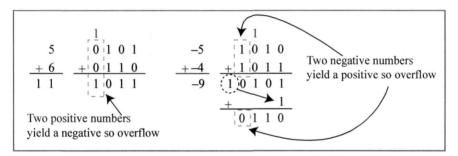

Figure 2.13   Two cases where an overflow has occurred. When two numbers of the same sign are added and they yield a different sign, an overflow has occurred and the result should be discarded.

Another way to state the overflow condition is that if the carry-in into the MSB position is different from the carry-out from the MSB position; then an overflow has occurred.

2.3.2.2 Subtraction in the 1's complement

Subtraction in the 1's complement is derived from addition. The process involves negating the number being subtracted and proceeding as addition. All rules introduced in addition apply. For completeness, an example of 1's complement subtraction is given in Figure 2.14.

Although the 1's complement simplifies arithmetic, the most commonly used negative number representation is the 2's complement because it has one 0 representation.

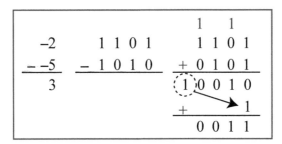

Figure 2.14   1's complement subtraction showing $-2 - (-5) = 3$. -2 is added to a negated -5.

2.3.3 The 2's complement

The *2's complement* representation is classified under the *radix complement* representation. Given an n-digit number X_1 in base r, we want to find another n-digit number X_2 that when added to X_1 will yield r^n. Writing in equation format yields

$$X_1 + X_2 = r^n$$

where X_1 is the given value and X_2 is its radix complement. If X_1 were a positive number, then X_2 is considered its negative counterpart, and vice versa. This representation works for all bases, but when the base $r = 2$, then the representation is called 2's complement.

By comparison, the 2's complement differs from the 1's complement by 1. To get a 2's complement number, find the 1's complement representation and then add 1.

The 2's complement number for a 4-bit number is depicted graphically in Figure 2.15. The 2's complement representation for a 4-bit number can code the numbers -8 through $+7$. Hence, n bits in the 2's complement can code numbers from $-(2^{n-1})$ to $+(2^{n-1} - 1)$. Notice that only one representation of 0 exists in this number representation. Also, notice that this representation always has one negative number without a positive number counterpart. In the 4-bit case, the negative number -8 ($= -2^{4-1=3}$) exists but the positive number $+8$ does not.

The weighted-sum approach for the 2's complement has the MSB weight as $-(2^{n-1})$, as shown in the equation:

$$D = -2^{n-1} \times a_{n-1} + 2^{n-2} \times a_{n-2} + \cdots + 2^0 a_0$$

An example is used to illustrate the conversion of a decimal number to a 2's complement number below for 59 and 67 using 8 bits.

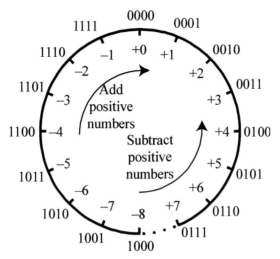

Figure 2.15 Graphically showing all numbers in a 4-bit 2's complement representation

Decimal number	59	67
Binary representation	111011_2	1000011_2
8-bit 2's complement	00111011_2	01000011_2
Negation (complement the bits and add 1)	$11000100_2 + 1_2 = 11000101_2$	$10111100_2 + 1_2 = 10111101_2$
Decimal verification of negation	$-2^{8-1} + 2^{8-2} + 2^{8-6} + 2^{8-8}$ $= -2^7 + 2^6 + 2^2 + 2^0 = -59$	$-2^{8-1} + 2^{8-3} + 2^{8-4} + 2^{8-5}$ $+ 2^{8-6} + 2^{8-8} = -2^7 + 2^5$ $+ 2^4 + 2^3 + 2^2 + 2^0 = -67$

The existence of only one representation of 0 in the 2's complement is attractive, but this comes with the side effect of an extra negative number. This number presents a special case. For example, take the 6-bit negative number 100000_2, which corresponds to $-2^5 = -32_{10}$. Proceeding with the negation of this number will yield $011111_2 + 1_2 = 100000_2$. Therefore, in the 2's complement negating 0 returns itself, and negating the extra negative number also returns itself.

Here's a summary of the properties of the 2's complement:

- Complement the bits and add 1 to change from a negative number to a positive number, and vice versa.
- If 0 is the MSB, then the number is positive; if 1 is the MSB, then the number is negative. Using this rule, we see that the representation for 0 is +0.

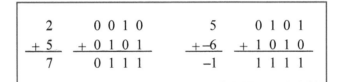

Figure 2.16   4-bit 2's complement addition showing $2 + 5 = 7$ and $5 + (-6) = -1$.

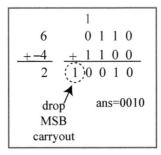

Figure 2.17   4-bit 2's complement addition showing $6 + (-4) = 2$. The carry-out from the MSB is dropped in 2's complement addition.

- There is only one representation of 0, and negating 0 returns 0
- An *n*-bit number in a 2's complement number lies between $-(2^{n-1})$ to $+(2^{n-1} - 1)$ in decimal.
- There is an extra negative number that is represented, and when negated this number returns itself.
- Positive number representations are identical to positive number representations in the 1's complement.

2.3.3.1 Addition in the 2's complement

Addition in the 2's complement is similar to that in the 1's complement with certain exceptions. Addition proceeds in the same manner as binary addition, as shown in Figure 2.16.

As in the 1's complement representation, sometimes the addition process yields a carry-out at the MSB. The carry-out is dealt with differently in the 2's complement. We ignore or drop the carry-out, as shown in Figure 2.17. Addition of 6 to –4 should yield 2. The addition yields 1 0010_2, but since we are dealing with 4-bit numbers, we ignore the carry-out bit from the MSB position to obtain the true result 0010_2, which is equivalent to 2 in decimal.

Overflow: As in the 1's complement, there is a possibility of overflow in 2's complement addition. Referring to the wheel in Figure 2.15, overflow occurs

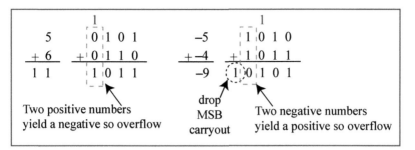

Figure 2.18   When two numbers of the same sign are added and they result in a number of a different sign, then an overflow has occurred.

when the dotted boundary between −8 and +7 is crossed in either direction during an addition process. The overflow rule is the same as that for the 1's complement case and is illustrated in Figure 2.18.

2.3.3.2 Subtraction in the 2's complement

2's complement numbers may be subtracted through binary subtraction, but since the order of numbers matters in the subtraction operation, we need to check the magnitude and sign of the given numbers before the operation of subtraction. Hence, using a binary-type subtraction does not yield any advantages to using the 2's complement representation, so a different method of subtraction is adopted. Subtraction in the 2's complement is accomplished by negation and addition as in the 1's complement case. The negation process, though, is split up separately into "complement and add 1." The add 1 is performed through a carry-in at the LSB after the complement is performed. *Note: The main advantage of this technique is that both addition and subtraction can be performed by the same hardware, which is a set of adder blocks, as will be discussed in later chapters.* An example of 2's complement subtraction is shown in Figure 2.19. Note that in Figure 2.19, "−5 complemented plus the carry-in of 1" = +5.

Although the extra negative number may not be negated successfully, it may still be used in arithmetic as long as an overflow does not occur. Two examples are shown in Figure 2.20 illustrate this point: (i) 5 is added to −8 to obtain −3, and (ii) −8 is subtracted from −6 to obtain 2. Both of these arithmetic operations involving −8 return the correct results since an overflow has not occurred. Try −8 − (−8) to see if it returns 0.

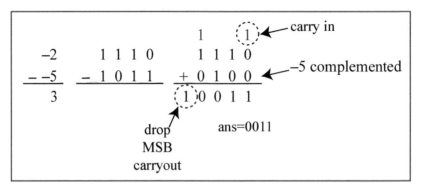

Figure 2.19   2's complement subtraction showing –2 – (–5) = 3. Subtraction involves negation of the subtracted number, which involves complementing the number and adding 1. 1 is added through a carry-in.

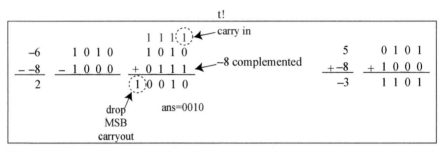

Figure 2.20   2's complement arithmetic with –8. Although –8 cannot be successfully negated, it may still be subtracted and added to other numbers as long as no overflow occurs.

2.4 Strings of Bits and Binary-Coded Decimal Representations

Binary strings like 11000010 can be compactly written in hexadecimal as C2. Yet if it is pertaining to room temperature in Fahrenheit, we, human beings, will have difficulty in converting the binary string into decimal values that we can easily recognize. Therefore, augmentation is needed to convert binary strings into binary-coded decimal (BCD) representations so that the hexadecimal number C2 can be represented as $2^7 + 2^6 + 2 = 194$ in the decimal system, which is written in BCD as 0001 1001 0100, where 0001 is the 4-bit binary representation for the decimal digit 1, 1001 is that for the decimal digit 9, and 0100 is that for the decimal digit 4. This specific 4-bit binary representation of a decimal number is called 8421 code, which is explained in the following section.

Input in hex	8421 BCD code	2421 BCD code	1357_9 BCD code	631-1 BCD code	84-2-1 BCD code	Excess-3 BCD code	Gray code
0	0000	0000	0000_0	0011	0000	0011	0000
1	0001	0001	1000_0	0010	0111	0100	0001
2	0010	0010	1000_1	0101	0110	0101	0011
3	0011	0011	0100_0	0111	0101	0110	0010
4	0100	0100	0100_1	0110	0100	0111	0110
5	0101	1011	0010_0	1001	1011	1000	1110
6	0110	1100	0010_1	1000	1010	1001	1010
7	0111	1101	0001_0	1010	1001	1010	1011
8	1000	1110	0001_1	1101	1000	1011	1001
9	1001	1111	0000_1	1100	1111	1100	1000
A	Invalid	Invalid	Invalid	Invalid	Invalid	Invalid	Invalid
...	...	...	...	...	...	...	...
F	Invalid	Invalid	Invalid	Invalid	Invalid	Invalid	Invalid

Figure 2.21   Examples of binary codes for decimal digit representation.

2.4.1 The 8421 BCD code

The 8421 BCD code follows the normal binary to decimal conversion, that is, it follows the following conversion equation:

$$\text{Decimal digit} = a_3 \times 2^3 + a_2 \times 2^2 + a_1 \times 2^1 + a_0 \times 2^0 = \sum_{i=0}^{3} a_i \times 2^i$$

where a_i is the binary digit of the ith place. For example, the decimal value of 0110 is 6 and is obtained as follows:

$$\text{Decimal digit} = 0 \times 2^3 + 1 \times 2^2 + 1 \times 2^1 + 0 \times 2^0 = 0 + 4 + 2 + 0 = 6$$

The naming of this coding technique as 8421 follows from the fact that 2^3 is 8, 2^2 is 4, 2^1 is 2, and 2^0 is 1; the name accounts for each place in the BCD number. Also, the table above (Figure 2.21) only accounts for values 0 through 9, the remaining possible binary inputs 1010 (A), 1011 (B), ..., 1111 (F) are invalid in BCD codes. Similar to the decimal system, after counting to 9, we start over but then add a 1 in front of the 0, that is, 0, 1, 2, ..., 9, 10, 11, ... Hence, the decimal number 15 in 8421 BCD is 0001 0101.

2.4.2 The 2421, 631-1, 84-2-1, and Excess-3 BCD codes

The 2421 BCD code is similar to the 8421 BCD code, but instead of 1111 corresponding to $8 + 4 + 2 + 1$, it corresponds to $2 + 4 + 2 + 1 = 9$. This code follows the same mechanism of decimal conversion for all digits. The code uses 0 through 4 in regular binary format, and then the codes for 5 through 9 are obtained by "reflecting and complementing" the bits of 4 to 0 in the code. The act of "reflection" or "mirroring" occurs about a hypothetical line that segregates the column $(0 - 9)$ into two parts: $(0 - 4)$ and $(5 - 9)$. The hypothetical mirror gives us a set of pairs of numbers: namely $(4, 5)$, $(3, 6)$, $(2, 7)$, $(1, 8)$ and $(0, 9)$ —the later number of a pair is the mirror image of the former one. For example, to obtain the 2421 representation of 5, we take the value of its mirror image $(4 \equiv 0100)$, and complement it to get 1011. Similarly, 0011 as 3 is complemented to 1100 and mirrored to the position corresponding to 2421 representation of 6, and so on.

The 631-1 BCD code works in a similar fashion, whereby the place values are 6, 3, 1, and –1. Hence to obtain the code for decimal 0, 0011 is used since this implies $(0 \times 6) + (0 \times 3) + (1 \times 1) + (1 \times -1) = 1 - 1 = 0$. The codes from decimal numbers 0 through 4 may be generated using the place-value system. The codes from 5 through 9 follow the place-value system as well as a self-complementing code sequence as seen in the 2421 BCD code (also known as the Aiken code). The decimal number 5 is 1001, which is the complemented form of the BCD representation of 4, 0110. As with the 2421 BCD code, the coded representation of the number 6 is the complement of 3, 7 is a complement of 2, and so on.

The 84-2-1 and Excess-3 BCD codes (also known as the Stibitz code) work in a similar way to the 631-1 and 2421 BCD codes as the values representing 5 through 9 are complemented versions of the values representing 0 through 4. Due to their nature, the codes discussed in this section are called *self-complementing codes*.

2.4.3 The biquinary code

1357_9 BCD (note the underscore symbol is not a minus sign) is a special type of 5-bit biquinary code used in Remington Rand vacuum tube computers, such as UNIVAC 60, built-in 1952. One vacuum tube would be ON at a time to represent one *quinary* bit for each of 1, 3, 5, and 7. The fifth *bi* bit represented 9 if the first four bits were not on. For even digits like 2, 4, 6, and 8, the fifth bi bit represented a value of 1, which was added to one of the first four bits, which was ON.

2.4.4 The gray code

In 1947, Bell Labs researcher Frank Gray developed a powerful coding technique, popularly known as the *Gray code*, which works by changing only one-bit value from one digit to another. For example, 0 is 0000, 1 is 0001, and 2 is 0011. Only 1 bit has changed from 0 to 1 and from 1 to 2. The changed bit is not random and follows a reflected pattern, as explained using Figure 2.21. From Figure 2.21, 0 is 0000, 1 is 0001, 2 is 0011, 3 is 0010, and 4 is 0110, whereas 5 through 9 are obtained by changing the MSB to 1 and reflecting the values of 4 through 0. For example, 4 is 0110 and 5 is 1110, and 3 is 0010, and 6 is 1010; only the MSB differs in these number pairs. The Gray code is a *reflective unit distance code* since only 1-bit changes between consecutive decimal representations, and it is a reflected version of the least significant bits, as demonstrated in Figure 2.21.

An application of the Gray code is the encoding of the angular position in a rotary encoder. Rotary encoders have been historically used in many applications, including the printer, fax machine, elevator, trackball mouse, and rotating radar platform. A rotary encoder can be realized in many forms, one of the implementations is the use of a mechanical disk with three distinct regions, as shown in Figure 2.22. Each region of the disk has black and white sectors, where black is translated as 1 and white is translated as 0. The disk in the figure is split up into 8 parts and encoded with 3 bits. By placing a sensor at a specific point, the angle of rotation can be measured with a certain degree of error. Figure 2.22 uses binary encoding in ascending order (counterclockwise direction) to identify the 8 parts.

The problem with this scheme is that when the encoder stops at a boundary, depending on the sensor, some noise or other errors might influence the readout of the encoder. Two possible error sources may be a shaky stop after rotation and a manufacturing defect where the code boundary is not distinct. For example, at the boundary between 011 and 100, the inner ring

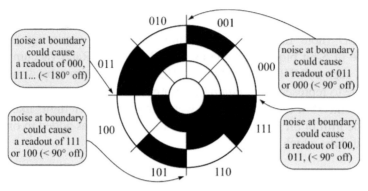

Figure 2.22   Mechanical disk encoder using a 3-bit binary code.

corresponding to the MSB may have a defect whereby the 0 region extends into the 100, region thereby creating an area in between that would read 000 instead of 100. If the encoder stops at this boundary and the sensor reads out 000, this constitutes about a 180° error on the position of the rotary encoder. If this system were used to determine the direction of a battleship's gun, the error may make the user think he or she is aiming in the opposite direction.

In the previous discussion, the boundary between 011 and 100 involves the change of 3 bits, that is, all bit positions change when transitioning. The maximum error that can be observed is a 180° error. If 2 bits changed, for example, at the boundary between 101 and 110, the maximum error would be 90°. If 1-bit changes, the maximum error is 45°.

The problematic boundaries where gross errors are possible are identified in Figure 2.22. To avoid this, most rotary encoders use codes such as the Gray code, which only has a 1-bit change in adjacent parts, as shown in Figure 2.23. If only 1 bit is changing between boundaries, the error is limited to the two adjacent regions. For example, referring to Figure 2.23, the boundary between 010 and 011 has the outer ring change. For instance, a manufacturing defect extended the outer ring, so the 010 region was a little bit larger than the 011 regions. The error in the readout of the angular position of the rotor resulting from this mechanical defect will be limited to 45°, whereas a similar defect in binary rotary encoder may result in an error of 180°.

A 45° precision is not that great. The 3-bit example is used to illustrate a concept, but actual rotary encoders use more than 3 bits and look a lot more like the 8-bit encoder (courtesy of parkermotion.com) in Figure 2.24.

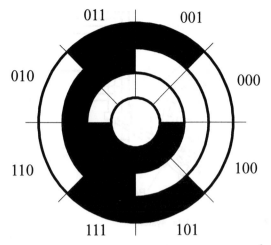

Figure 2.23 Mechanical disk encoder using the Gray code.

Figure 2.24 Mechanical disk encoder using an 8-bit Gray code.

Thus Gray code, accompanied by a larger-bit encoding system can reduce the margin of error to a significant extent.

Gray Encoded Electrical Signals Using Optical Techniques

Figure 2.25 illustrates how a rotating mechanical disk generates Gray coded binary signals by using a light source called a light-emitting diode (LED). The light is collimated by an optical system to pass through the rotary disk according to black and white coding as shown in Figure 2.23. Phototransistors having high selectivity are placed at the other end of the disk

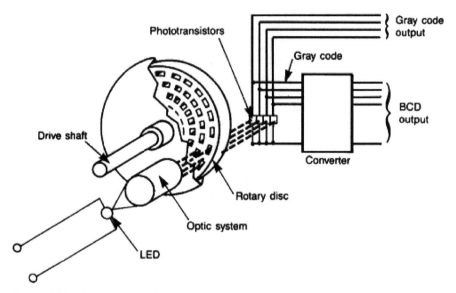

Figure 2.25   Gray encoded digital signal generation by using an optical source (Courtesy: PLCDev that manufactures programmable logic controller devices.)

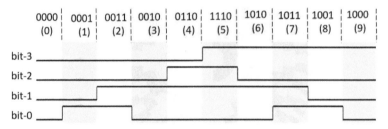

Figure 2.26   Electrical signals for the Gray code generated by the phototransistors by converting the optical beams transmitted through the rotary disk.

to sense the light generated by LED and convert optical signals into Gray encoded digital signals, as shown in Figure 2.26.

Generally, the ON and OFF times for digital signals are unequal when they are generated in this way through the transduction of optical signals into an electrical signal. If the ON time period is lesser than the OFF time period of the transduced digital signal, an adoption of Gray encoding is preferred to conventional binary encoding, as the former can prevent any temporary spurious output that could be resulted from the ON/OFF period mismatch. Note that the successive Gray code differs by only one position in the code. For example, when the disk changes from an angular position corresponding to numeral 7_{10} (i.e., 0111_2) to the numeral 8_{10} (i.e., 1000_2), the disk will rotate by one change of color as 7_{10} is 1011 in Gray code while 8_{10} is 1001. Hence, only one phototransistor will turn OFF, while three other phototransistors will remain ON. This is true for all other numerals changing to its next numeral value, as by construction Gray code will differ by one position in the code.

However, if we use a mechanical disk, which is encoded according to binary code in Figure 2.22, then a change of 7_{10} to 8_{10} will require one phototransistor (MSB) to turn ON, while the other three phototransistors must turn OFF. As the turn ON time is lower than turn OFF time, the digital signals will undergo a spurious transition when all phototransistors are turned ON: $0111 \rightarrow 1111 \rightarrow 1000$. Therefore, in Figure 2.25 a separate converter unit is included in the diagram to transform Gray encoded digital signals to normal BCD digital signals.

2.4.5 BCD summary

As with the complement systems, the BCD system has addition and subtraction that are not shown here. The addition and subtraction are essentially decimal addition and subtraction. Some computer processors like the IBM Power6 have been designed to work with BCD number system processing. The BCD system has the following advantages over binary complement systems:

- Many fractional representations, such as decimal 0.2, have an infinite place-value representation in binary (0.001100110011 . . .) due to never-ending repeating digits, but they have a finite place value in BCD (0.0010). Due to this, there are fewer rounding errors when using BCD compared to binary.
- Since decimal numbers are used in everyday situations, especially in financial situations where dollars are converted to cents and vice versa, the multiplication and division by 10 are more easily accomplished with BCD than with binary.

- The alignment of two decimal numbers (e.g., $1.7 + 98.18$) is a simple, exact shift.
- Converting digits to a display is easily accomplished in BCD. For example, the microwave display has a seven-segment display that uses BCD representation. This is because each digit on the display is mapped to a BCD number; a binary number representation would use too much complex logic.

Most of the advantages of the BCD system stem from the ease of interpretation of a decimal number by a human. The disadvantages of BCD, on the other hand, stem from the fact that it warrants bigger logic circuits as opposed to that in binary. The disadvantages of BCD include the following:

- Some arithmetic operations are more complex to implement: adders require extra logic to perform correctly. About 15% to 20% more circuitry is required in BCD addition compared to binary. Multiplication requires the use of algorithms more complex than binary multiplication.
- Standard BCD requires 4 bits per digit, about 20% more space than a binary encoding. Compression methods that reduce standard BCD to 3 digits in 10 bits reduce the storage overhead but misalign storage with the common 8-bit byte boundaries common on hardware today (compatibility problem). This results in slower BCD implementations on these systems.
- BCD implementations are not widely supported by all processors, especially those in embedded systems, such as a router; hence, BCD operations will run slower than binary operations.

2.5 More on Number Systems

Numbers can be classified into number systems, which have been historically extended as the complexity of the uses of numbers increased. *Natural numbers* came about to be used first and are 1, 2, 3, ... and so on. Traditionally, 0 was not considered a number; however, in the 19th century, 0 was also added to the set of natural numbers. In some sources, when 0 is included in the natural numbers, the new set is referred to as *whole numbers*. If negative numbers are added to the set of whole numbers, the new set is referred to as *integers*. Integers are –3, –2, –1, 0, 1, 2, 3, and so on.

The next number system is called *rational numbers*. Rational numbers are represented as a fraction of two numbers, where the numerator is an integer and the denominator is a positive integer. Some rational numbers are

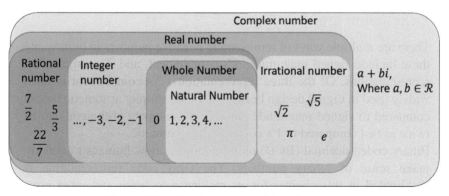

Figure 2.27 Number system and its subsystems.

$\frac{1}{2}, \frac{3}{8}, \frac{-1}{5}, \ldots$ and so on. However, there exist some numbers that cannot be represented as the ratio of two integers; these are referred to as *irrational numbers*. A famous example of irrational numbers is the number π, whose value is $3.1415926\ldots$, and the digits after the decimal point do not repeat or end.

Rational and irrational numbers together make up the *real number system*. Real numbers are a superset including all of the number sets we have discussed so far. The way the number sets relate to each other is visualized in Figure 2.27. You can see that natural numbers are a subset of whole numbers, which are a subset of integers. Integers are a subset of rational numbers. Rational and irrational numbers are both a subset of real numbers.

A *complex number* is represented as $a + bi$, where a and b are real numbers and i is the solution to $\sqrt{-1}$. The number a is referred to as the real part, and the number b is referred to as the imaginary part. Complex numbers are a superset of real numbers because they can be thought as 2D numbers having a real and an imaginary component, whereas real numbers only have a real component. Therefore, if a complex number has a 0 imaginary component, it is a real number, and if it has a 0 real component, it is a purely imaginary number. Complex number examples are $5 + 7i$, $-3 + 6.2i$, $-8 - \sqrt{7}i$, $-5, 3i$, and so on. Complex numbers are widely used in science and engineering in numerical constructs that cannot be represented by just using real numbers.

2.6 Conclusion and Key Points

1. There are multiple ways of representing negative numbers in binary, and these include signed magnitude, 1's complement, and 2's complement representations. Of the three representations, 2's complement is most widely used in digital design because it yields simpler arithmetic blocks compared to signed magnitude, and 2's complement arithmetic can be twice as fast compared to 1's complement arithmetic.

2. Binary coded decimal (BCD) representations allow humans to quickly make sense of binary numbers. The Gray code representation is highlighted in this chapter for its resilience to noise and other error compared to other BCD representations.

Problems on Number Systems

These are merely a few sample practice problems. Click on the publisher's website to find a large set of problems.

1. *Number conversion*:

   (a) Convert 137.32_{10} to hexadecimal and then to binary.
   (b) Convert $EAD.123_{16}$ to a base-4 number.
   (c) Convert $A75.11_{11}$ to a base-5 number.
   (d) Convert $EAC.784_{16}$ to a base-7 number.

2. *Binary addition and subtraction*:

   (a) Add (−12) and (−16) by using 1's and 2's complement representations.
   (b) Subtract (−11) from (−15) by using 1's and 2's complement representations.
   (c) Add (9) to (−14) by using 1's and 2's complement representations.
   (d) Subtract (−7) from (13) by using 1's and 2's complement representations.
   (e) If you are a computer designer, will you *prefer* the 1's complement to the 2's complement for the subtraction operation? Explain your reasons.

3. The following 6-bit words represent signed binary integers. Give in ordinary decimal form the number represented by each word interpreted for signed-magnitude code and 2's complement code.

Number word	Signed magnitude	2's complement
011 111	_____	_____
101 110	_____	_____
110 011	_____	_____
010 100	_____	_____

4. The following 8-bit hexadecimal words represent signed binary integers. Give in ordinary decimal form the number represented by each word interpreted for signed-magnitude code and 2's complement code.

Number word	Signed magnitude	2's complement
AE	_____	_____
D5	_____	_____
6E	_____	_____
3C	_____	_____

5. A computer has a word length of 8 bits (including the sign). If 2's complement is used to represent negative numbers, what range of integers can be stored in the computer? Repeat the above if the 1's complement is used to represent negative numbers.

6. The base-10 fraction 1/6 is converted to base 3 and the result is rounded off to 4 digits. Show the number in base 3.

7. (a) What is the number of bits required for the BCD representation of an arbitrary n-digit decimal integer?

   (b) Represent the decimal numbers 0 through 9 in 842-1 and 442-1 BCD codes

8. *Troubleshooting of digital hardware*: Assume a microprocessor uses a 16-bit databus connected to external memory devices to store unsigned binary data. To troubleshoot faults, a service engineer needs to monitor the contents of the bus as the microprocessor stores information into the memory devices. The service engineer uses a logic probe that picks up binary data from the databus and displays the contents of the databus in decimal format. Assume that the probe senses the following set of binary data in hexadecimal format: 1AE7, 5B2F, 0C21, DEAD, and FFFF. The logic probe uses an internal circuit to convert the above 16-bit information into $4n$-bit BCD data. What is the minimum value of n? Write the corresponding BCD codes for each of the above six hexadecimal data values, if the logic probe uses the Stibitz code for displaying BCD data. Repeat the process for the case when the logic probe uses the Aiken code. Note that the maximum and minimum

unsigned binary data that the microprocessor stores are FFFF and 0000, respectively.

9. *Innovative digital gadget*: A fancy digital clock uses 24 light-emitting diodes (LEDs) organized into 4 rows and 6 columns to display the time of the day. Assume that each column represents a BCD code to represent the decimal value of the time and the top LED in each column represents the MSB. If an LED is switched on, it is indicated by a red color dot. On the contrary, if the LED is switched off, it is indicated by a light-gray color dot. By drawing separate 4 × 6 colored LED matrices, display the following times of the day in 24-hour format: 7:13:45 A.M., 11:59:59 A.M., 12:47:23 P.M., 9:38:19 P.M., and 11:48:27 P.M.

10. *Reliable computing by encoding data and address bits*: The IBM 650 (built-in 1953) computer used biquinary-coded decimal data and addresses to incorporate an error-checking scheme that ensured reliable operation. A biquinary code uses 7 bits comprising two *bi* bits, 0 and 5, and five *quinary* bits, 0, 1, 2, 3, and 4. Therefore, a biquinary code has the positional weights of 05-01234. For example, the decimal value of 6 is represented as 01-01000 and the biquinary code of decimal value of 3 is 10-00010. Write the biquinary code for all 10 decimal values and show that there are exactly two 1s and five 0s in the code, no matter what the decimal numeral is. Therefore, if any of the 7 bits inverts randomly due to noise or some other sources of error, the computer will be able to identify erroneous data.

11. *Mechanical rotary encoders for car stereo volume control*: Incremental mechanical encoders typically have two quadrature outputs, A and B, which are 90° out of phase. They are used to track angular motion and determine position and velocity. An encoder provides cyclical quadrature outputs when it is rotated. Write two separate tables showing the encoding schemes for clockwise rotation and counterclockwise rotation of the encoder. You need to indicate the binary values of the quadrature outputs, A and B, corresponding to phases 1, 2, 3, and 4 in each table. If state 1 represents a high-voltage level and state 0 a low-voltage level, draw the waveforms for A and B corresponding to various phases in both clockwise and counterclockwise rotations.

12. *Mechanical rotary encoder*: Figures 2.22 and 2.23 in Section 2.4.4 show encoding schemes for mechanical disks with binary code and Gray code, respectively. Assuming black represents a signal value of 1 and white represents 0, draw the waveforms of 4 bits for each type of mechanical disk as it rotates by 360°.

13. *Angular rotation encoded using 4-bit natural binary and Gray codes*: Assume that the decimal code 0 represents an angular rotation in the $0°$–$22.5°$ range, 1 represents the $22.5°$–$45°$ range, and so on. Write a table with four columns, where the first column shows the decimal code from 0 to 15, the second column shows the corresponding rotation range in degrees, while the third and fourth columns indicate the binary and Gray codes, respectively.

14. *The Mayan number system*: The Mayan civilization (2000 BC–1539 AD) used a variant of the positional number system, where the radix was 20, and it used a combination of only two symbols: a dot ($\cdot$) and a dash ($-$). For example, 2 is represented by two dots, $\cdot\cdot$; 4 by four dots, $\cdot\cdot\cdot\cdot$; 5 by a dash, $-$; 9 by four dots over a dash, $\cdot\cdot\cdot\cdot/-$; and 18 by three dots over three dashes, stacked vertically, $\cdot\cdot\cdot/-/-/-$. The Mayan mathematicians symbolized the concept of zero by using a symbol of a shell, $\emptyset$. The Mayans represented their numbers vertically with the LSB at the bottom (20^0) and the MSB at the top of the vertical stack of symbols. This vertical positional number representation as opposed to the horizontal positional number system (discussed in Section 2.1.1) used by other civilizations. Show the following decimal numbers in the Mayan number system by stacking the symbols: 30, 45, 536, 1776, and 23,998.

15. *Non-Boolean Gray code*: The ternary or 3-ary Gray code uses three numeral symbols, namely 0, 1, and 2. The n-ary Gray code with k digits is known as the (n,k)-ary Gray code. For example, a $(3,2)$-ary Gray code comprises a sequence of $9 = 3^2$ code words: 00, 01, 02, 12, 10, 11, 21, 22, 20. Write a complete table comprising $27 = 3^3$ ternary code words and the corresponding $(3,3)$-Gray code words.

16. *Base-20 Number System*: Base-20, which is also known as Vigesimal number system, uses 0–9 and A–H, J, and K as symbols. The alphabet I is skipped as it can be misunderstood as numeral 1. Represent the reciprocal of 2, 3, 5, 7, and 11 (all decimal prime numbers) as their corresponding base-20 numbers, showing up to 6 places after the radix point.

17. *Sexagesimal Number System*: In base-60 number system, one uses a comma to denote the different positions of a digit and a semicolon to denote the fraction part of the number as Ptolemy had used to denote the value of pi (π) in base-60 system as 3;8, 30. Note that $\pi = 3.141\,666$ can be obtained from the base 60 value as $3 + 8 \times 60^{-1} + 30 \times 60^{-2}$.

i. Show that 1;24, 51, 10 in base-60 system represents the value of the diagonal of a unit square. Apply Pythagoras Theorem to calculate the length of the diagonal of a unit square, which is an irrational number.

ii. The Earth revolves around the Sun in 365.24579 days as calculated by ancient astronomers applying the base-60 systems. Represent the above decimal number in the base-60 system.

iii. Represent binary numbers: 0.1_2, 0.001_2, and 0.000001_2 into their corresponding base-60 numbers.

Boolean Algebra and Logic Gates

This chapter introduces the basic theorems of Boolean algebra using switching circuits. Later some properties and characteristics of Boolean representations are discussed and logic gate implementations are used to show and develop schematic representations of circuits. Complementary metal-oxide semiconductor (CMOS) gate implementations are discussed to wrap up the chapter.

Terms introduced in this chapter: Huntington postulates, duality, self-dual, bubble, equivalent gates

Competency Objectives: At the end of this chapter, you will be able to:

1. Use the Huntington postulates and DeMorgan's Laws to rearrange and/or minimize Boolean expressions.
2. Learn how to represent Boolean expressions in the form of algebra as well as that of switching logic.
3. Learn that determining a dual of a Boolean expression results in a true statement.

4. Learn the concept of completeness and what set of Boolean operators constitute a complete set and which logic gates form a complete set.
5. Evaluate basic CMOS logic gates.

3.1 Motivation

The renowned British mathematician and logician George Boole (1815–1864) first proposed Boolean algebra in 1847 to solve mathematical problems involving deductive logic. Claude Shannon (1916–2001), who is widely acclaimed as the father of *information theory*, was the first to apply Boolean algebra to the design of switching networks while working on his master's thesis at the Massachusetts Institute of Technology (MIT) in 1937. Essentially, Shannon translated Boolean algebra to switching algebra, hence completing the adaptation of Boolean algebra to designing digital logic systems. Boolean algebra has proven indispensable to understanding and designing digital computers; therefore, this chapter will strive to provide a sufficient background on the topic.

3.2 Huntington Postulates

Here is a list of properties associated with Boolean algebra, and later, these properties will be proven using switching circuits:

1. Operation on at least two distinct logic elements in some domain B.
2. Property of closure, that is, $X + Y \in B$ and $X \cdot Y \in B$ given that $X, Y \in B$.
3. Commutative property, that is, $X + Y = Y + X$ and $X \cdot Y = Y \cdot X$.
4. Associative property, that is, $X + (Y + Z) = (X + Y) + Z$ and $X(Y \cdot Z) = (X \cdot Y) \cdot Z$.
5. Identity property, that is, $X + 0 = X$ and $X \cdot 1 = X$.
6. Distributive property, that is, $(X + Y) \cdot Z = (X \cdot Z) + (Y \cdot Z)$ and $(X \cdot Y) + Z = (X + Z) \cdot (Y + Z)$.
7. Existence of a complement: for all $X \in B$, there exists X'(or $\overline{X}$) whereby $X + X' = 1$ and $X \cdot X' = 0$.

Properties 1 through 3 and 5 through 7 are *Huntington postulates*, named after the American mathematician Edward Huntington (1874–1952). Property 4, the associative property, is not listed as a postulate since it can be derived from the commutative property.

We will define domain B to contain the set $\{0, 1\}$. In switching algebra, we will represent a value of 0 as an open switch (open circuit) and a value of 1 as a closed switch (closed circuit).

There are three operations in Boolean algebra, namely the AND operation, denoted with "·"; the OR operation, denoted with "+"; and the inversion operation, denoted with an apostrophe "'". These three operations can be used to make any conceivable Boolean function. In switching circuits, the operation "+" corresponds to a *parallel* connection, while the operation "·" corresponds to a *series* connection.

3.3 Basic Theorems of Boolean Algebra

Reiterating, 0 and 1 are Boolean values in binary representation corresponding to the OFF and ON states of mechanical switches and the LOW and HIGH values of output logic gates.

3.3.1 Basic postulates with 0 and 1

When $X = 1$, the mechanical switch is in the ON position, and when $X = 0$, the mechanical switch is in the OFF position. The combination of Boolean operators $(+, \cdot,$ and $')$ and values (0 and 1) produces relational properties, as those summarized later. The input in this case is a battery, and the Boolean variable X is used in different switch configurations in order to affect the output (light bulb). A radiant bulb denotes a lamp that is ON, a white bulb denotes a lamp that is OFF, and a black bulb denotes a lamp in a state directly dependent on the value of X. In the case of the black bulb, if $X = 1$, the bulb is ON, and if $X = 0$, the bulb is OFF.

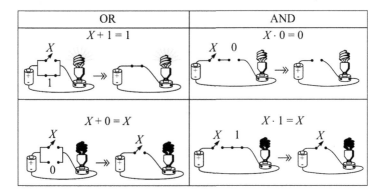

3.3.2 Idempotent laws

Any Boolean value that is ANDed or ORed with itself returns the same value.

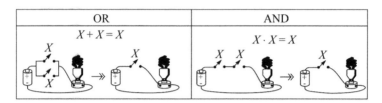

OR	AND
$X + X = X$	$X \cdot X = X$

3.3.3 The law of involution

The symbolic representation for an inversion operation is

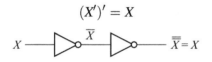

This symbol is referred to as the *inverter symbol* or a *NOT gate*. When 0 is at its input, it returns a 1 at its output; when 1 is at its input, it returns 0 at its output.

$$(X')' = X$$

$$X \longrightarrow \overline{X} \longrightarrow \overline{\overline{X}} = X$$

3.3.4 Complementarity laws

Using switching networks with two states, it is clear to see that in the OR case, when one switch is closed, the other switch is opened; hence, the lamp will always be ON. In the AND case, since both switches are in series, the lamp will be OFF.

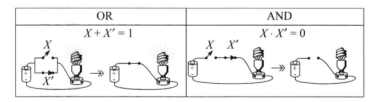

OR	AND
$X + X' = 1$	$X \cdot X' = 0$

3.3.5 Commutative laws

Switching the order of variables in an OR operation or in an AND operation does not change the Boolean function.

OR	AND
$X + Y = Y + X$	$X \cdot Y = Y \cdot X$

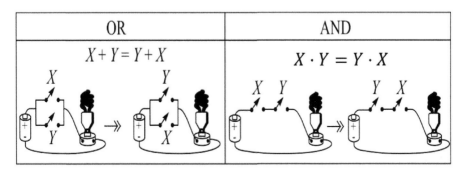

3.3.6 Associative laws

The figures are simplified a bit, that is, the battery and lamp are no longer shown. But it is assumed they are attached, and the Boolean variables are used to determine if the lamp is ON or OFF.

OR	AND
$(X + Y) + Z = X + (Y + Z)$	$(X \cdot Y) \cdot Z = X \cdot (Y \cdot Z)$

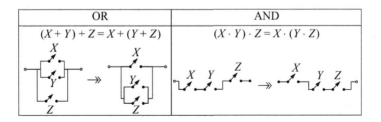

3.3.7 Distributive laws

A clear difference between algebra and Boolean algebra is in the application of the distributive law. The OR and AND operations distribute, as shown. In algebra, normally "+" does not distribute as shown, but this is legal in Boolean algebra.

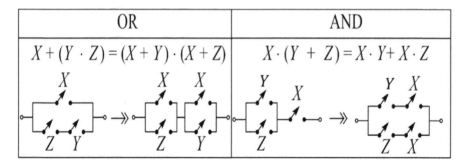

3.3.8 DeMorgan's laws

DeMorgan's laws are summarized as:

$$(X+Y)' = X' \cdots Y' \quad | \quad (X \cdots Y)' = X' + Y'$$

Self-study: A quick, helpful exercise is to use switching networks to confirm these relations. Hint: Like in the previous examples, assume a default state for each switch and stick with the convention throughout.

Both of the DeMorgan's laws can be generalized to greater than two variables following these equations:

$$(X_1 + X_2 + X_3 + \ldots + X_N)' \quad | \quad (X_1 \cdot X_2 \cdot X_3 \cdot \ldots \cdot X_N)'$$
$$= X_1' \cdot X_2' \cdot X_3' \cdot \ldots \cdot X_N' \quad | \quad = X_1' + X_2' + X_3' + \ldots + X_N'$$

Example 1: Prove the consensus theorems.
Using both Boolean algebra and switching algebra, prove the consensus theorems given here:

OR	AND
$X \cdot Y + Y \cdot Z + X' \cdot Z = X \cdot Y + X' \cdot Z$	$(X+Y) \cdot (Y + Z) \cdot (X' + Z)$
	$= (X + Y) \cdot (X' + Z)$

Boolean algebra approach for the OR case

Procedure: Always start with the given equation; every transformation done must adhere to the aforementioned laws and postulates. Try not to skip steps, and always write down which law is applied in each step.

Step 0: $X \cdot Y + Y \cdot Z + X' \cdot Z$	Given LHS of consensus theorem
Step 1: $X \cdot Y + (X + X') \cdot Y \cdot Z + X' \cdot Z$	Complementarity law
Step 2: $X \cdot Y + X \cdot Y \cdot Z + X' \cdot Y \cdot Z + X' \cdot Z$	Distributive law
Step 3: $X \cdot Y \cdot (1 + Z) + X' \cdot Y \cdot Z + X' \cdot Z$	Distributive law
Step 4: $X \cdot Y \cdot 1 + X' \cdot Y \cdot Z + X' \cdot Z$	Basic postulates
Step 5: $X \cdot Y + X' \cdot Y \cdot Z + X' \cdot Z$	Basic postulates
Step 6: $X \cdot Y + Y \cdot X' \cdot Z + X' \cdot Z$	Commutative law
Step 7: $X \cdot Y + (1 + Y) \cdot X' \cdot Z$	Distributive law
Step 8: $X \cdot Y + 1 \cdot X' \cdot Z$	Basic postulates
Step 9: $X \cdot Y + X' \cdot Z$	Basic postulates

Common practice: Every step was shown in this example, and even the commutative law had a step of its own. Usually, the commutative law is easy to recognize and is not necessary to be shown from step to step. The same applies for the application of basic postulates that result in 0 or 1. It is not necessary to write down the 1.

Switching algebra approach for the OR case

Procedure: Convert the given equation to switch format and apply transformations corresponding to the switching interpretation of Boolean algebra laws and postulates.

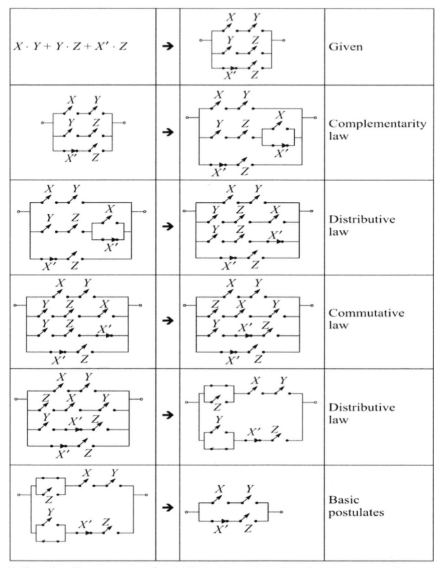

$X \cdot Y + Y \cdot Z + X' \cdot Z$	→		Given
	→		Complementarity law
	→		Distributive law
	→		Commutative law
	→		Distributive law
	→		Basic postulates

Self-study: Try to prove the AND relationship for the consensus theorem, following a similar method outlined for the OR relationship.

3.4 Duality

To form the dual of a Boolean expression, replace AND with OR, OR with AND, 1 with 0, and 0 with 1. Do not modify variables or complements. When done properly, the resultant Boolean expression should express a true statement. For example, from the basic postulates we see the following expressions are dual:

$$X + 1 = 1 \quad \overset{\text{dual}}{\Longrightarrow} \quad X \cdot 0 = 0$$
$$X + 0 = X \quad \overset{\text{dual}}{\Longrightarrow} \quad X \cdot 1 = X$$

Here are two more examples of duality. Again, do not modify variables or their complements:

$$(X + Y) \cdot \left(1 + X\overline{Y}\right) \quad \overset{\text{dual}}{\Longrightarrow} \quad (X \cdot Y) + \left(0 \cdot (X + \overline{Y})\right)$$
$$(X + Z) \cdot XYZ \cdot \left(0 + \overline{X} + \overline{Y}\right) \quad \overset{\text{dual}}{\Longrightarrow} \quad X \cdot Z + X + Y + Z + \left(1 \cdot \overline{X} \cdot \overline{Y}\right)$$

Duality claim 1: For any function it can be shown that $\overline{f}(X_1, X_2, \ldots, X_N) = f^D\left(\overline{X}_1, \overline{X}_2, \ldots \overline{X}_N\right)$.
   For example, given $f(X, Y, Z) = X + Y + Z$.
   Then $\overline{f}(X, Y, Z) = \overline{X + Y + Z}$ and $f^D(X, Y, Z) = X \cdot Y \cdot Z$.
   Applying DeMorgan's law to $\overline{f}(X, Y, Z)$ yields $\overline{f}(X, Y, Z) = \overline{X} \cdot \overline{Y} \cdot \overline{Z}$.

$$\therefore \overline{f}(X, Y, Z) = f^D(\overline{X}, \overline{Y}, \overline{Z})$$

Duality claim 2: Normally, for a general function, $f \neq f^D$.
   In our previous example, $f(X, Y, Z) = X + Y + Z$ and $f^D(X, Y, Z) = X \cdot Y \cdot Z$; therefore we can see that $f \neq f^D$.

Duality claim 3: If $f = f^D$ then f is called a *self-dual function*.
   For example, given $f(X, Y, Z) = XY + YZ + XZ$.

$$
\begin{array}{ll}
f^D(X, Y, Z) = (X + Y) \cdot (Y + Z) \cdot (X + Z) & \\
f^D(X, Y, Z) = (XY + XZ + YY + YZ) \cdot (X + Z) & \text{Distributive law} \\
f^D(X, Y, Z) = XXY + XXZ + XYY + XYZ & \text{Distributive law} \\
\quad + XYZ + XZZ + YYZ + YZZ & \\
f^D(X, Y, Z) = XY + XZ + XY + XYZ + XZ + YZ + YZ & \text{Idempotent law} \\
f^D(X, Y, Z) = XY + XZ + XYZ + YZ & \text{Idempotent law} \\
f^D(X, Y, Z) = XY \cdot (1 + Z) + XZ + YZ & \text{Distributive law} \\
f^D(X, Y, Z) = XY + XZ + YZ & \text{Basic postulates} \\
\quad \therefore f = f^D &
\end{array}
$$

3.4.1 What you should know from duality

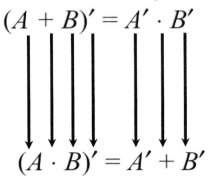

$$(A + B)' = A' \cdot B'$$

$$(A \cdot B)' = A' + B'$$

The principle of duality tells us that if, in a Boolean equation, we interchange the AND and OR operators and interchange 0s and 1s, then the resultant Boolean equation is also true.

Caution: It does not mean that both the Boolean and its dual are equal; the concept of duality is a way of representing another true statement.
To generate the dual of a Boolean function:

(1) Interchange 0 and 1.
(2) Interchange AND and OR.
(3) Keep the NOTs as-is.
(4) Keep the literals as-is. (trues and complements are not interchanged).

If a given function's dual expression returns the same logic function, the given function is said to be self-dual.

3.5 Logic Gates for Implementation of Boolean Networks

3.5.1 AND, OR, and NOT

In switching network representations, the value of the Boolean variables (X and Y) is used to determine the state (ON or OFF) of a lamp. In digital circuits, the Boolean variables X and Y represent inputs to a system, and the state of the lamp represents another Boolean variable F. An expression can be written for the state of the lamp as a function of X and Y, that is, $F(X,Y)$. Applying this concept of Boolean functions, the OR operation can be represented schematically as an OR gate and the AND operation can be represented schematically as an AND gate.

Figure 3.1 shows the representation of three types of logic gates, two of them are two-input logic gates and the remaining one is a one-input logic

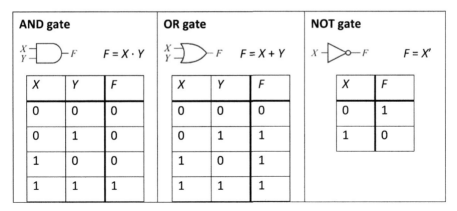

AND gate			OR gate			NOT gate	
X Y >— F $F = X \cdot Y$			X Y >— F $F = X + Y$			X —>o— F $F = X'$	

X	Y	F	X	Y	F	X	F
0	0	0	0	0	0	0	1
0	1	0	0	1	1	1	0
1	0	0	1	0	1		
1	1	1	1	1	1		

Figure 3.1   Two-input AND, OR, and NOT gate symbols and truth tables defining each function.

gate. The AND gate returns an output of 1 (the lamp is ON) whenever both inputs are 1. The OR gate returns an output of 1 whenever any of the inputs is 1. The NOT gate or inverter performs the inversion operation.

Notice that in Figure 3.1, the truth tables follow an enumeration in ascending order with respect to the variables X and Y. A truth table contains all possible input combinations. To avoid mistakes of duplicates or omission, listing all combinations in a logical order is advised. The first row can be seen as row 0 because the binary number XY is 00, the second row can be seen as row 1 since XY is 01, the third row is row 2 since XY is 10, and last row is row 3 since XY is 11. In this case, X is seen as the most significant bit (MSB) and Y is seen as the least significant bit (LSB) as we count from 0 to 3.

For the one-variable case, the NOT gate, only two lines exist. For the two-variable case, four lines exist. Since there are only two possible values $\{0,1\}$ each Boolean variable can take, and there are two variables, then the number of permutations that can be made are $2 \times 2 = 4$; hence, there are 4 lines in the truth table. If there were three variables, then it would be $2 \times 2 \times 2 = 8$. The relationship between the number of variables n_v and the number of rows n_r in the truth table therefore follows as $n_r = 2^{n_v}$.

3.5.2 Implicant gates

Given a gate G that takes two inputs, a and b, there would be four different output possibilities based on the input cases: 00, 01, 10, and 11. Since each output case can take on two values, 0 or 1, and there are four positions to fill

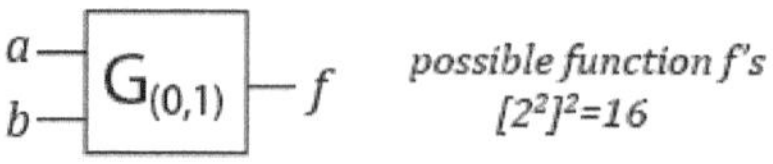

$$a - \boxed{G_{(0,1)}} - f \qquad \text{possible function } f\text{'s} \\ b - \qquad\qquad\qquad [2^2]^2 = 16$$

a	b	f_0	f_1	f_2	f_3	AND	XNOR	XOR	$a \to b$	$b \to a$	ab'	...
0	0	0	0	1	1	0	1	0	1	1	0	
0	1	0	1	0	1	0	0	1	1	0	0	
1	0	0	1	0	1	0	0	1	0	1	1	
1	1	0	1	0	0	1	1	0	1	1	0	

Figure 3.2   Number of possible Boolean functions for a two-input gate is 16, and some function outputs are listed.

corresponding to the four cases, the number of distinct output combinations for gate G turns out to be $2^4 = 16$.

Figure 3.2 shows a table with f_1 implementing the OR function, f_2 the NOR, up until the implicant gates. Implicant gates are $a \to b$ (referred to as a implies b) and $b \to a$ (referred to as b implies a).

$$a \to b \text{ can be written as } ==> a' + b.$$

$$b \to a \text{ can be written as } ==> a + b'$$

This relationship is important for logicians, for example, in the logical proof where a and b are as follows:

$a \to$ all creatures are mortal,
$b \to$ man is a creature,
$\therefore a \to b$ man is mortal.

3.5.3 Other logic gates

AND, OR, and NOT are not the only gates used in digital design. The use of two variables suggests that 16 different logic functions can be obtained, as discussed in Section 3.5.2. All 16 functions' truth table is shown in Figure 3.3; X and Y are inputs, and $F\#$ are outputs. We have already seen $F1$, the AND function, and $F7$, the OR function. The other functions' descriptions are summarized in Figure 3.4.

To round off the list of the most visible Boolean gates you will come across; see Figure 3.5. The two input gates correspond to $F14$, $F8$, $F6$, and $F9$ in Figure 3.4.

X	Y	F0	F1	F2	F3	F4	F5	F6	F7	F8	F9	F10	F11	F12	F13	F14	F15
0	0	0	0	0	0	0	0	0	0	1	1	1	1	1	1	1	1
0	1	0	0	0	0	1	1	1	1	0	0	0	0	1	1	1	1
1	0	0	0	1	1	0	0	1	1	0	0	1	1	0	0	1	1
1	1	0	1	0	1	0	1	0	1	0	1	0	1	0	1	0	1

Figure 3.3 Truth table for all 16 logic functions that can be made from two Boolean variables.

Boolean function	Operator symbol	Name	Comments
$F0 = 0$	—	NULL	Constant 0
$F1 = XY$	$X \cdot Y$	AND	X and Y
$F2 = XY'$	X/Y	Inhibition	X but not Y
$F3 = X$		Transfer	X
$F4 = X'Y$	Y/X	Inhibition	Y but not X
$F5 = Y$	Y	Transfer	Y
$F6 = XY' + X'Y$	$X \oplus Y$	Exclusive OR (XOR)	X is not equal to Y
$F7 = X + Y$	$X+Y$	OR	X or Y
$F8 = (X + Y)'$	$X \downarrow Y$	NOR	Not OR
$F9 = XY + X'Y'$	$X \odot Y$	Exclusive NOR (XNOR)	X equals Y
$F10 = Y'$	Y'	Complement	Not Y
$F11 = X + Y'$	$X \subset Y$	Implication	If Y then X
$F12 = X'$	X'	Complement	Not X
$F13 = X' + Y$	$X \supset Y$	Implication	If X then Y
$F14 = (XY)'$	$X \uparrow Y$	NAND	Not AND
$F15 = 1$	—	Identity	Constant 1

Figure 3.4 Boolean expressions for all 16 functions.

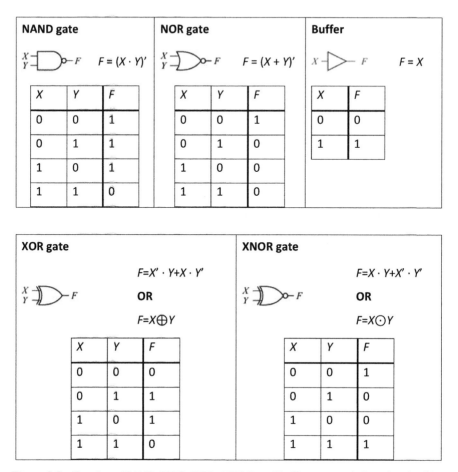

Figure 3.5   Two-input NAND, NOR, XOR, XNOR, and buffer gate symbols and truth tables defining each function.

The NAND truth table looks like the output of AND put through an inverter, hence the name NAND (Not AND). Similarly, the same logic applies for the NOR gate (Not OR). The XOR gate output is 1 whenever X is different from Y, and the XNOR gate output is 1 whenever both X and Y are the same value. XNOR is the inverted form of the XOR and vice versa. The symbols of NAND, NOR, XNOR, and NOT gates all have a *bubble* at the output. This visual distinction signifies an inverted output, that is, the original function is evaluated and the output is inverted. For example, the NOR symbol is essentially hinting that it is an OR function followed by an inversion.

The practice of using bubbles to signify or identify inverted outputs hints at the possibility of using the same for inverted inputs. Bubbles at inputs allow for compact drawings of equivalent gates, as discussed next.

3.5.4 Equivalent gates

Equivalent gates are gates that perform the same functions. For example, an OR gate connected to the input of an inverter is equivalent to a NOR gate.

Figure 3.6 gives various examples of equivalent gates using the bubble concept. The first row provides gate examples that are equivalent to the XOR function, the second row provides examples equivalent to the XNOR function, the third row provides an equivalent gate example for the NAND gate, and the last row provides an equivalent gate for the NOR gate.

The NAND and NOR equivalent gates are directly from DeMorgan's laws: $(X + Y)' = X' \cdot Y'$ and $(X \cdot Y)' = X' + Y'$. For the XOR and XNOR cases, the equivalent gates can be proved either using truth tables or using Boolean algebra.

Self-study: Prove the following expressions are indeed equivalent:
$$X \oplus Y = \overline{X} \oplus \overline{Y} = \overline{X \oplus \overline{Y}} = \overline{\overline{X} \oplus Y}$$

and

$$X \oplus Y = \overline{\overline{X} \oplus \overline{Y}} = X \oplus \overline{Y} = \overline{X} \oplus Y$$

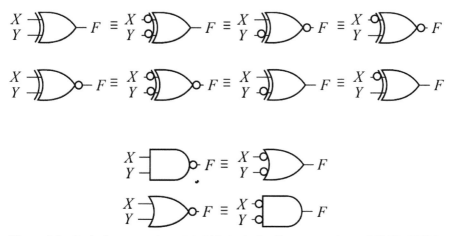

Figure 3.6 Equivalent gates through bubble logic for (from top to bottom) XOR, XNOR, NAND, and NOR.

Example 2: Simplify the following schematic.

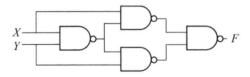

Procedure: Start by identifying the expressions at the intermediate nodes and writing the expression for F.

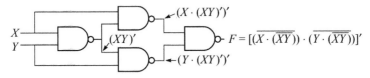

Then simplify the obtained expression using Boolean algebra.

$$F = \left[\overline{\left(X \cdot \overline{(XY)}\right) \cdot \left(Y \cdot \overline{(XY)}\right)}\right]$$

$F = \overline{X \cdot \left(\overline{XY}\right)} + \overline{Y \cdot \left(\overline{XY}\right)}$	DeMorgan's law
$F = \overline{X} \cdot \left(\overline{X} + \overline{Y}\right) + \overline{Y} \cdot \left(\overline{X} + \overline{Y}\right)$	DeMorgan's law
$F = \overline{X} \cdot X + \overline{X} \cdot \overline{Y} + \overline{Y} \cdot X + \overline{Y} \cdot Y$	Distributive law
$F = \overline{X} \cdot \overline{Y} + Y \cdot \overline{X}$	Complementarity law
$F = X \oplus Y$	

Hence

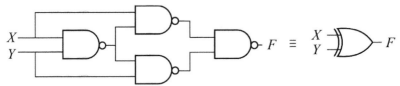

Self-study: Simplify the following schematic.

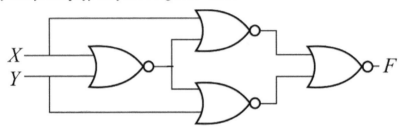

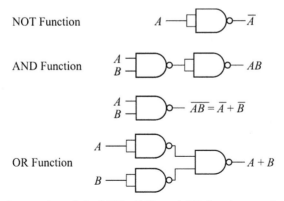

NOT Function

AND Function

OR Function

Figure 3.7 Implementation of the NOT, AND, and OR functions to show the NAND is functionally complete.

3.5.5 Concept of completeness

A functionally complete set of Boolean operators $\{\cdots\}$ is one that can express all possible truth tables by combining members of the set $\{\cdots\}$ into a Boolean expression. For example, the set $\{$AND, OR, NOT$\}$ is a complete set. To test the completeness of any other set, it suffices to show that the members of the given set can be combined to implement the AND, OR, and NOT operation. As an example, we will show that the NAND gate is complete by implementing AND, OR, and NOT functions using only NAND gates. This means that any digital circuit configuration can be implemented with just NAND gates!

Figure 3.7 shows the implementation of the AND, OR, and NOT functions using only NAND gates. The NOT function is realized by tying both inputs of the NAND gate together. The AND function is realized by using the NOT gate after a NAND operation.

To implement OR function, we recall that DeMorgan's Law provides a mechanism of conversion from the NAND operation on the given signals to the OR operation on the complement of the signals. Hence OR can be realized by inverting the input signals, followed by NAND operation.

3.6 CMOS Gates

Gate realizations differ depending on the implementing technology. For example, under the hood, a CMOS realization of a gate will be different from a transistor–transistor logic (TTL) realization of the same gate. From the perspective of the gate level, what is important is the gate's functionality. From the technology level, for example, CMOS, what is important is the transistor implementation of the gate specifications. For example, these specifications may include the operating voltage range, low-signal and high-signal margins, etc. This section will introduce CMOS transistors, notably the P-type metal-oxide semiconductor (PMOS) and the N-type metal-oxide semiconductor (NMOS[1]) (Figure 3.8), and will show how these transistors can be used to build familiar logic gates.

The description and modeling of transistor behavior can become detailed. Since this textbook is concerned with the digital behavior of CMOS transistors; a basic transistor model is presented. An NMOS is a transistor that is good at passing logic 0 from its source to its drain, while a PMOS is a transistor that is good at passing logic 1 from its source to its drain. The use of this concept will be clearer when we talk about transistor-level gate implementation. With the proposed model, transistors are essentially viewed as voltage-controlled switches: When the voltage at the gate of the NMOS is logic 1, then the switch is closed; otherwise, the switch is open. From bubble logic, the PMOS works in the opposite way: When the voltage at the gate of the PMOS is logic 0, the switch is closed; otherwise it is open. When the switch is closed, the source is directly connected to the drain. Using these concepts about the NMOS and PMOS, logic gates like the inverter (Figure 3.9) can be designed in a CMOS.

The CMOS inverter is made with a PMOS and an NMOS, as shown in Figure 3.9. A symmetric design is employed whereby the top half of the circuit is implemented with PMOS transistors and the bottom part of the

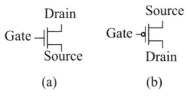

Figure 3.8   Symbolic representation of (a) an NMOS transistor and (b) a PMOS transistor.

---

[1]Since students who are not in the Electrical or Computer Engineering program are not accustomed to drawing digital circuits using the 3-terminal transistor symbols, as shown in Figure 3.8, a simpler way to draw an NMOS switch is to make a thin rectangle with two terminals (drain and source) at its two ends and a third terminal (gate) that controls the switching action. A PMOS switch is differentiated from an NMOS switch by adding a bubble at the control input. See Problem #8 in the exercise at the end.

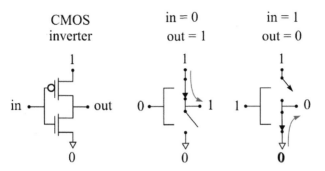

Figure 3.9 CMOS inverter showing that PMOS and NMOS transistors can be viewed as switches.

circuit is implemented with NMOS transistors. The PMOS switch is used to pass a 1, while the NMOS is used to pass a 0. The input **in** is connected to both the NMOS and PMOS transistors. Whenever **in** is 0, the PMOS transistor will act like a closed switch and will pass 1 to the output. Whenever **in** is 1, the NMOS transistor will act like a closed switch and will pass 0 to the output.

The NAND gate is realized in a CMOS, as presented in Figure 3.10. When either input A or B is 0, there is a direct connection of 1 to the output. For the output to be 0, both A and B need to be 1. Using switching circuits, some input combinations are shown in Figure 3.10.

The NOR gate is realized in a CMOS, as presented in Figure 3.11. As with the NAND gate, some input combinations are also shown.

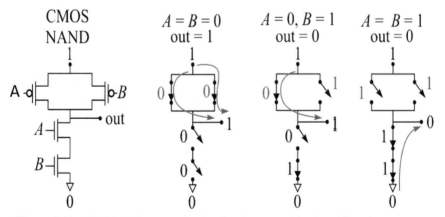

Figure 3.10 CMOS NAND gate showing how input combinations affect the output.

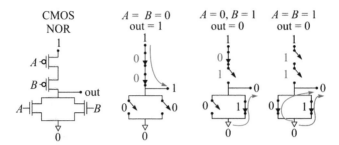

Figure 3.11 CMOS NOR gate showing how input combinations determine the output.

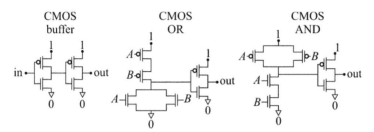

Figure 3.12 Implementation of noninverting gates (buffer, OR, AND) in a static CMOS.

In a CMOS, noninverting gates use more transistors than inverting gates. The buffer, OR, and AND gate implementations are shown in Figure 3.12. Noninverting gates are obtained by using inverters at the output of the inverting gates. Therefore, if possible use NAND and NOR as opposed to AND and OR in order to reduce the number of transistors used to implement logic functions.

3.7 General Complementary Switching Network

Figure 3.13 shows the implementation of a Boolean function F(A,B) by an alternative representation of the general complementary switching network. In this representation, a switch is denoted by a rectangular box with a line protruding out from the middle. The protruded line is the control terminal of the switch. A control terminal without a bubble symbol (with a bubble symbol) indicates that the switch is activated by asserting 1 (asserting 0) at the control input.

These switches can be constructed with any arbitrary technology and the concept of Boolean digital systems can be applied. Note that the upper switching network pulls up the output to 1 if A is not equal to B, when

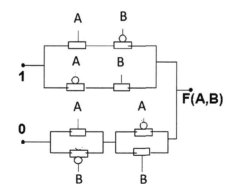

Figure 3.13 Implementation of a Boolean function in arbitrary technology with 3-terminal switches

the lower switching network, which is dual of the upper switching network, disconnects the output from 0. If A = B, then the output is pulled down to the value 0 through lower switching network, while the upper network remains disconnected.

Such pull-up and pull-down type complementary switching networks are needed if the switches excites with parasitic memory elements such as capacitors. If the arbitrary technology representing the switches does not have to charge-up parasitic memory elements such as capacitor or inductor, then the output function can be implemented merely using the upper pull-up network alone. In that case, when A = B, the upper network will disconnect the output from 1, and it will be assumed that F(A,B) = 0 when A = B.

3.8 Conclusion

This chapter introduced the properties of Boolean algebra and defined Huntington postulates, the concept of duality, and how to find the dual of a given function. Later, an introduction to schematic representations was presented with the introduction of logic gates, specifically AND, OR, NOT, buffer, NAND, NOR, XOR, and XNOR. The idea of building equivalent gates was introduced with examples of how to build XOR gates out of NAND gates. In addition, bubble logic was used to extend the concept of equivalent gates through inverted inputs and inverted outputs. Finally, CMOS logic implementation of the basic gates, excluding XOR and XNOR, was briefly introduced.

3.9 Key Points

1. Boolean algebra and switching algebra can be used to rearrange Boolean expressions by applying the Huntington postulates and DeMorgan's Laws.
2. The dual of a Boolean expression results in a true statement. To determine the dual of a Boolean expression, replace AND with OR, OR with AND, 1 with 0, and 0 with 1. Do not modify variables or complements.
3. NAND gates can be used to implement all Boolean functions.
4. CMOS circuits use fewer transistors in implementing NAND and NOR gates as opposed to AND and OR gates.

Problems on Boolean Algebra

1. Using Boolean algebra show the following equalities:

   (a) $(X \oplus Y)' = X \odot \bar{Y} = \bar{X} \oplus Y = XY + \bar{X} \cdot \bar{Y}$
   (b) $(X \oplus Y) \oplus Z = X \oplus (Y \oplus Z) = X \oplus Y \oplus Z$
   (c) $X \oplus Y = X' \oplus Y' = (X' \oplus Y)' = (X \oplus Y')'$
   (d) $X\bar{Y} + XYZ + \bar{X}Z = (\bar{X}\bar{Z} + Y\bar{Z})'$

   Proofs using a truth table are not acceptable.

2. Using Boolean algebra determine if the following equalities are true or false. Show each step.

   (a) $XY + Y'Z + X'Z = XY + Z$
   (b) $ab'c' + a'bd + ab\,(c \odot d) + (a \oplus b)\,d = ac' + d$

   Proofs using a truth table are not acceptable.

3. Find the duals of the following:

   (a) $(0 + a)\,.b + (1.b)$
   (b) $a.\bar{c}.d + b.c$
   (c) $((a + \bar{c})\,.1)\,.(b + d.0) + c.\bar{b}$

4. Implement the Boolean expressions given using corresponding digital logic gates.

   (a) $f\,(a, b, c) = a\overline{(b + c)} + \bar{a}b$
   (b) $f\,(a, b, c) = \overline{ab}(a \oplus c)$
   (c) $f\,(a, b, c, d) = \overline{((ab\bar{d})\,(\bar{b} + c))}(\bar{a}c)$

5. Apply Boolean algebra to write the minimized Boolean function f in terms of the inputs a, b, c, and d, where the function is represented in the following table. Implement the minimized Boolean function using AND, OR and NOT digital logic gates.

a	b	c	d	f
0	0	0	0	0
0	0	0	1	1
0	0	1	0	0
0	0	1	1	1
0	1	0	0	0
0	1	0	1	0
0	1	1	0	1
0	1	1	1	1
1	0	0	0	0
1	0	0	1	0
1	0	1	0	0
1	0	1	1	0
1	1	0	0	1
1	1	0	1	0
1	1	1	0	0
1	1	1	1	0

6. Find the most simplified version of the output expression for the gate circuit given using Boolean algebra.

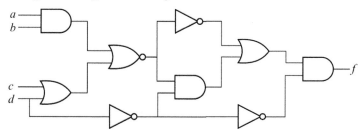

7. Which of the following expressions are equivalent to $f(a, b, c, d) = \bar{a}d + \bar{a}b\bar{c} + bcd + a\bar{c}d + a\bar{b}c\bar{d}$.

A. $f(a, b, c, d) = \overline{a}\overline{d} + \overline{a}bd + bcd + a\overline{b}\overline{d} + \overline{c}\overline{d}$
B. $f(a, b, c, d) = \overline{a}\overline{d} + \overline{a}bd + bcd + c\overline{d}$
C. $f(a, b, c, d) = \overline{b}\overline{d} + \overline{c}\overline{d} + \overline{a}b + bcd$
D. $f(a, b, c, d) = \overline{b}\overline{d} + \overline{c}\overline{d} + \overline{a}b + cd$
E. None of the above

8. Find the values of the inputs (a,b,c,d) of the switch implementation such that its output Z_1 is exactly identical to the output Z_2 of the indicated logic implementation. Note that a and b are connected to NMOS type switch and c and d are connected to PMOS type switch, respectively.

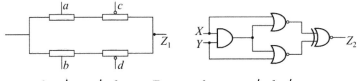

A. $a = x$, $b = y'$, $c = y'$, $d = x$ B. $a = x$, $b = y$, $c = y'$, $d = x'$
C. $a = x$, $b = y'$, $c = y$, $d = x'$ D. $a = x'$, $b = y'$, $c = x$, $d = y$
E. None of the above

9. In the following diagram, what are relationships between $F1$, $F2$, $F3$, and $F4$?

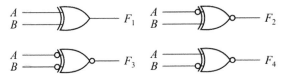

A. $F_1 = A \oplus B$, $F_2 = F_3 = F_4 = A \odot B$ B. $F_1 = F_2 = A \oplus B$, $F_3 = F_4 = A \odot B$
C. $F_1 = F_2 = F_3 = A \oplus B$, $F_4 = A \odot B$ D. $F_1 = F_2 = F_3 = F_4 = A \oplus B$
E. None of the above

10. Given the following four-gated networks, which of the statements is true?

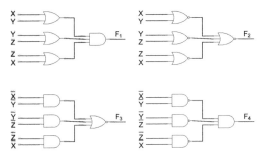

A. $F_1 = F_2$ and $F_3 = F_4$, but $F_1 \neq F_3$.
B. $F_1 = F_2$ and $F_3 = F_4$, but $F_2 \neq F_4$.
C. $F_1 = F_3$ and $F_2 = F_4$, but $F_1 \neq F_2$.
D. $F_1 = F_4$ and $F_2 = F_3$, but $F_2 \neq F_4$.
E. $F_1 = F_2$ and $F_3 = F_4$, but $F_2 = F_4$.

11. A complementary switch circuit generates the logic function $z(x, y) = \overline{x} + y$. The pull-up circuit (C) that connects z to 1 is given, but the pull-down circuit ($\overline{C}$) that connects z to 0 is not specified. Which of the following circuits should be used for $\overline{C}$?

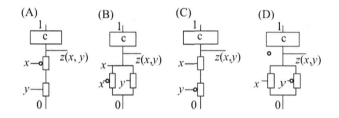

Comment: In Figure B, fix the diagram by removing x at the left middle. In Figure D, remove o as that is not part of the problem)

12. *Expansion of the truth table with an unknown input* (X): In the truth table discussed in Section 3.5, the binary inputs of a logic gate are either 0 or 1. The output of the truth table relates to the 1 (true) and 0 (false) associations of the inputs. However, if an input is unknown, that is, it can be 0 or 1, it is denoted by X. Note that in a two-input AND gate, if an input is X, the output is X if the other input is 1, while the output is 0 if the other input is 0. Similarly, in a two-input OR gate, if an input is X, the output is X if the other input is 0, while the output is 1 if the other input is 1. The truth table with unknown input can be compactly represented in matrix format, as shown below for two-input AND and two-input OR gates. Using the same matrix format, write the truth table with unknown input (X) for two-input NAND, two-input NOR, two-input XOR, and two-input XNOR gates. You will later find the use of the extended truth table in the design and analysis of large Boolean networks.

13. *Ternary (also called three-valued, trinary, and trivalent) logic*: Boolean algebra can also be applied to Boolean variables if they assume more than two states. In ternary logic, a Boolean variable has three states, denoted by 0, 1, and 2. In terms of a truth table, 1 can be considered as truth and 2 can be considered as false, while the third state 0 can be considered as "maybe," that is, perhaps true or perhaps false. Thus

in a ternary NOT gate, an input of 1 produces an output of 2, and vice versa, while a 0 input produces an output of 0. Using the same concept, write the truth table for two-input NAND, two-input NOR, two-input XOR, and two-input XNOR gates in ternary logic. Look into the US Patent 4,107,549 (1978) for ternary logic gates that can be constructed by modifying binary CMOS gates.

14. *Testing of faulty Boolean gates*: In real-world digital circuits that are fabricated by CMOS process technology, often a gate input or output may be shorted to ground (V_{SS}) or supply (V_{DD}) wires. The real challenge is for designers to apply appropriate test inputs to check whether the logic gates have any faults. When a gate input or output is shorted to V_{SS}, it is known to be having a stuck-at 0 defect. When a gate input or output is shorted to V_{DD}, it is known to be having a stuck-at 1 defect. When the expected input variable is 1, but its observed value is 0 as it is stuck-at 0, we define a new symbol D to express the state of the input in the truth table. On the contrary, when the expected input variable is 0, but its observed value is 1 as it is stuck-at 1, we define a new symbol D' to express the state of the input in the truth table. In the following diagram, the truth table of a NOT gate is shown with D and D' input states in addition to unknown and fault-free states. Using the above concept, write the truth table for two-input NAND, two-input NOR, two-input XOR, and two-input XNOR gates to include faulty states D and D' along with unknown and fault-free states.

NOT gate Truth Table

Input	Output
0	1
1	0
X	X
D	D'
D'	D

15. *Test generation for faulty Boolean networks*: Given the following network of Boolean gates, assume that the input $n2$ of G4 is stuck-at 0, as shown in the diagram. To observe the fault at the output (Z), the inputs A, B, C, and D must be appropriately set. Explain your answer by discussing how the D value at $n2$ can be propagated as D' at Z.

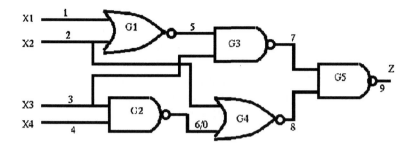

16. *Implementation of a Boolean network*: Using AND, OR, and NOT gates implement the Boolean expression $Z(X, Y, A, B, C, D) = X'Y'A + X'YB + XY'C + XYD$. Note that depending on the values of inputs X and Y, which are called *select* inputs, the output (Z) assumes the values of inputs A, B, C, and D, which are called *data* inputs. In other words, the above Boolean network is a multiplexer that selects one of the four data inputs by asserting the select inputs.

17. *Dual of the Boolean network*:

    (a) Write the Boolean function (Z) for the minority gate.

    (b) Applying Boolean laws show that $Z = Z^D$, that is, Z is a self-dual function.

    (c) Draw the Boolean network of Z^D by using NOR gates only.

18. *Implementation of a Boolean function using a CMOS*: Implement the following Boolean expressions by using a PMOS for a pull-up network (PUN) and an NMOS for a pull-down network (PDN):

    (a) $Z_1(A,B,C) = (A'+B')(B'+C')(C'+A')$

    (b) $Z_2(A,B,C) = A'B' + B'C' + C'A'$

    Draw truth tables for Z_1 and Z_2. If they are identical, then apply Boolean algebra on the expression of Z_1 and show that it reduces to the expression of Z_2.

19. *Boolean algebra to reduce CMOS transistors*:

    (a) Given a three-variable Boolean function, $Z(A,B,C) = A'BC + AB'C +ABC' + ABC$, where A is the MSB and C is the LSB, draw a schematic of CMOS transistors that will implement the function such that the PMOS transistors will set $Z = 1$ for four input patterns and NMOS transistors will set $Z = 0$ for the other four input patterns. Distinguish which input patterns set $Z = 1$ and which of them set $Z = 0$.

(b) Now apply Boolean algebra laws to reduce the above function to $Z(A,B,C) = AB + BC + CA$, where A is the MSB and C is the LSB. Draw a schematic of CMOS transistors that will implement the function $Z(A,B,C)$. Verify that the PMOS transistors will set $Z = 1$ for four input patterns in the previous case and NMOS transistors will set $Z = 0$ for the other four input patterns. Find the saving of number of PMOS and NMOS transistors.

Timing Diagrams

This chapter provides an introduction to timing, delay, and functional decomposition. Timing is important in circuit design, referring to the microprocessor shown in Figure 4.1, because information may need to travel long distances, as evidenced by the long interconnects or wires (lighter colors). The dark sections of the microprocessor are mostly cache memory, and as you can see, memory takes up most of the microprocessor chip area. Reading and writing to memory is not accomplished instantaneously and must be timed correctly to avoid errors.

Micro-Processor

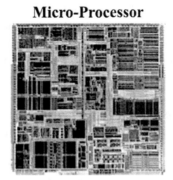

Figure 4.1   Microprocessor.

Intel's Core i7 processors boast over 700 million transistors on a die area of fewer than 300 mm^2. Careful design is critical!

Terms introduced in this chapter: gate delay, rise delay, fall delay, interconnect delay, fall time, rise time, propagation delay, reconvergence, static-1 hazard, static-0 hazard, path delay, critical path, and critical delay

Competency Objectives: At the end of this chapter, you will be able to:

1. Learn the concept of gate delay and interconnect delay; differentiate between rise delay and fall delay and know which CMOS transistors contribute to rise and fall delays.
2. Analyze timing diagrams or waveforms at various nodes of circuits.
3. Build an oscillator using inverter gates.
4. Determine the critical path of a circuit.

4.1 Notion of Timing Delay in a Circuit

Two types of circuit delays often cause timing problems in a digital circuit: gate delay and interconnect delay. *Gate delay* is dependent on the actual realization of the gate. For example, in the complementary metal-oxide-semiconductor (CMOS) two-input NAND gate depicted in Figure 4.2, the information at the output Z is stored in the form of electric charges on an intrinsic capacitor because of the structure of the transistors comprising the gate. Depending on the logic inputs at A and B and the *present value stored at* Z, the storage capacitor may either discharge through the two N-type metal-oxide-semiconductor (NMOS) transistors (Q1 and Q3) or charge up through one or both of the P-type metal-oxide-semiconductor (PMOS) transistors (Q2 and Q4). Input changes that result in charging the capacitor at output Z through the PMOS devices will undergo a delay dubbed *rise delay* since the output capacitor is changing from a discharged (0 V) state to a charged (V_{DD}) state. Input changes that result in discharging the capacitor at output Z through the NMOS devices are dubbed *fall delay* since the capacitor is going from a charged state to a discharged one. In essence, *gate delays* come about because a finite amount of time is required for gate output(s) to respond to changes at its input(s).

[Comment: Need to take permission for the microphotograph]

Interconnect delay on the other hand stems from nonideal characteristics (Figure 4.2b) of wires and connectors that are used to connect transistors to configure into gates. *Interconnect delays* are nonideal effects associated

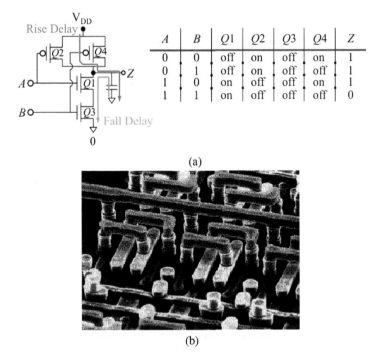

A	B	Q1	Q2	Q3	Q4	Z
0	0	off	on	off	on	1
0	1	off	on	off	off	1
1	0	on	off	off	on	1
1	1	on	off	off	off	0

(a)

(b)

Figure 4.2 (a) CMOS NAND gate showing paths for charging and discharging the storage capacitor and (b) zoomed-in SEM picture of the microprocessor showing interconnects are not perfectly straight or composed of one metal piece.

with parasitic capacitances and resistances of wires. The scanning electron microscopy (SEM) photo shows that interconnects are made up of different shapes and materials that influence wire properties. Due to these nonidealities, voltage change along the length of a wire is not an instantaneous feat. During timing analysis, interconnect and gate delays may be lumped together at the gate output, as shown in Figure 4.3 (the AND gate with gate + wire delay = 3 nanoseconds, which is written in acronym as ns). In this example, we are given that at time $t = 0$, the gate inputs, a and b, are 0 and 1, respectively, and the output y is 0. At time $t = 1$, input a switches from 0 to 1, but this switch is not realized at the output y instantaneously. Due to the gate + wire delay of 3 ns, the switch in the output y from 0 to 1 is realized at time $t = 4$ instead of at time $t = 1$ when input a switches. Following the waveform, at time $t = 5$, input b switches from 1 to 0, but this input change is only realized at the output Y at time $t = 8$ when y switches from 1 to 0.

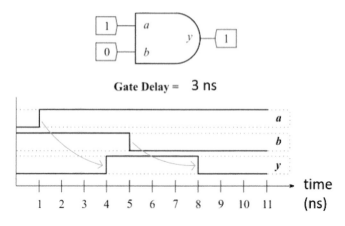

Figure 4.3   (Top) AND gate showing the values of inputs and output between times t = 5 and t = 8. (Bottom) A timing diagram showing the effect of gate delay on output response time due to change in inputs.

4.2 Definition of Propagation Delay

For the convenience of a more rigorous analysis of timing, we are going to define some terminologies, such as rise and fall delay, propagation delay, etc. Rise (fall) delay is the time difference between a change at input X at its 50% mark to a rising (falling) edge at output Y's 50% mark, as shown in Figure 4.4. The rise delay in the Equation 4.1 is denoted as t_{pLH} to signify the output signal going from low to high, and the fall delay is denoted as t_{pHL} to signify the output signal changing from high to low. The rise and fall delay are not necessarily equal. We typically take an average of both the delays and call it the *propagation delay* (t_{d}).

$$t_{\mathrm{d}} = \frac{t_{\mathrm{pLH}} + t_{\mathrm{pHL}}}{2} \tag{4.1}$$

Figure 4.4 also has a table summary of sample chips. The Logic Family column describes the technology such as complementary metal-oxide-semiconductor (CMOS) and transistor-transistor logic (TTL) used to make the chip. The Function column describes what kind of logic function the chip performs. The propagation delays vary due to process variation within each technology, so typical values are listed on specification sheets. Sometimes typical values are not shown, but maximum values of propagation delay are required. Every integrated circuit (IC) gate you buy will have specifications, as shown.

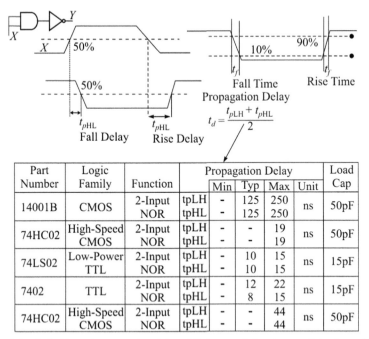

Part Number	Logic Family	Function		Propagation Delay				Load Cap
				Min	Typ	Max	Unit	
14001B	CMOS	2-Input NOR	tpLH	-	125	250	ns	50pF
			tpHL	-	125	250		
74HC02	High-Speed CMOS	2-Input NOR	tpLH	-	-	19	ns	50pF
			tpHL	-	-	19		
74LS02	Low-Power TTL	2-Input NOR	tpLH	-	10	15	ns	15pF
			tpHL	-	10	15		
7402	TTL	2-Input NOR	tpLH	-	12	22	ns	15pF
			tpHL	-	8	15		
74HC02	High-Speed CMOS	2-Input NOR	tpLH	-	-	44	ns	50pF
			tpHL	-	-	44		

Figure 4.4 Definition of propagation delay, fall time, and rise time explained graphically.

In addition to propagation delay, there are metrics called *fall time* and *rise time*, which correspond to a measure of how quickly a signal changes from one logic level to another. Fall time is the measure of the time difference at the falling edge of a signal between 90% and 10% of its maximum value, while rise time is the measure of the time difference at the rising edge of a signal between the 10% mark and the 90% mark. The 10% and 90% metrics are used as opposed to 0% and 100% because signals do not look clean and neat in real life: This is especially true for multi-Gigahertz signals; 0% and 100% would be hard to determine with confidence under noisy signal conditions.

In conclusion, propagation delay is a delay between the input and the output signals and is measured as the duration between the 50% signal levels for both input and output signals. The fall time and rise time are transient properties seen just at a single signal, for example, just at the output. Determination of propagation delay involves input and an output signal, whereas rise time and fall time involves a single, particular signal. *Do not confuse rise time with rise delay or fall time with fall delay.*

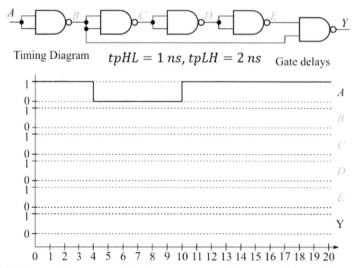

Figure 4.5   Timing diagram providing the parameters of the given problem and the waveform of the given input *A*.

4.3 Timing Diagrams of a Gated Logic

With the definitions of rise time, fall time, rise delay, fall delay, and propagation delay covered, we are now ready to analyze the timing of a gated logic in more detail. The circuit schematic in Figure 4.5 shows five NAND gates in a serial configuration. The first four NAND gates, which determine outputs *B*, *C*, *D*, and *E*, are configured to be used as inverters, while the last gate, which determines output *Y*, is used as a NAND gate. The fall delays, denoted as $t_{pHL} = 1$ ns, are the same for all the gates, and the rise delays, denoted as $t_{pLH} = 2$ ns, are the same for all the gates. The input signal *A* is given in the timing diagram as switching from Logic 1 to Logic 0 at time $t = 4$ ns and later switching from Logic 0 to Logic 1 at time $t = 10$ ns.

The change in input signal *A* will directly affect the signal at node *B*, as shown in Figure 4.6. We are given the value of *A* starting at time $t = 0$ ns. Before the time of $t = 0$ ns state of A is UNKNOWN. If fall delay is given as 1 ns, state of B will become KNOWN after 1 ns time elapses since A takes a KNOWN state. If *A* were 1 before time $t = 0$ ns, then *B* would be 0 in the shaded region; but if *A* were previously 0 and switches to 1 at time $t = 0$ ns, then *B* would start from 1 and then switch to 0 at time $t = 1$ ns. The value of *B* is only certain after time $t = 1$ ns.

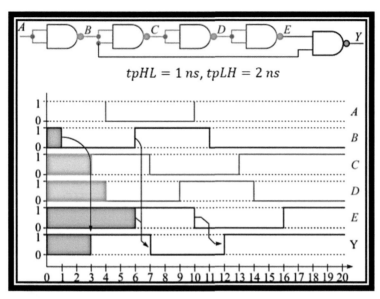

Figure 4.6   Timing diagram isolating and showing the change in output Y due to its inputs B and E.

The analysis continues for node B whereby the change in A from 1 to 0 at time $t = 4$ ns is realized in B at time $t = 6$ ns because the rise delay for the gates is given as 2 ns. This edgewise analysis is followed throughout to realize the wave pattern for signal B. After the waveform for B is completed, then C is analyzed in a similar way, and the process continues until output Y. Output Y is a bit more complex because the two signals that Y depends on, B and E, change at different times due to a change in input A. The change in A from 1 to 0 at time $t = 4$ ns causes B to change to 1 at time $t = 6$ ns. The change in B at time $t = 6$ ns, coupled with the current value of E, causes Y to change at time $t = 7$ ns. The change in A at time $t = 4$ ns propagates to E at time $t = 10$ ns, so when E changes at this time, this change is realized in Y after 2 ns at time $t = 12$ ns. The complete timing diagram is shown in Figure 4.7, delineating the cause-and-effect relationship between the change in input A at time $t = 4$ ns and the intermediate nodes of the circuit until the output Y changes.

In the same manner as how rise delay was defined in Figure 4.4, rise delay can be defined for any string of gates or complex logic as long as the input and output are clearly defined. Rise delay from input A to output Y can be calculated from the timing diagram of Figure 4.7 as $12 - 4 = 8$ ns. Note that

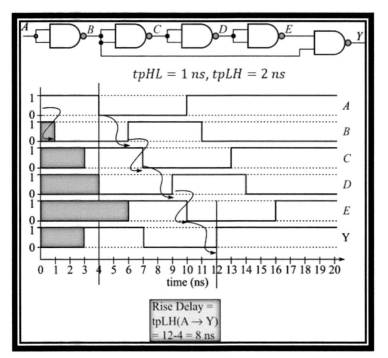

Figure 4.7   Complete timing diagram defining the rise delay of the entire chain of NAND gates.

the rise delay corresponds to the input change that causes the output to switch from 0 to 1: *output rises*! This is portrayed in Figure 4.7.

4.4 The Ring Oscillator: Good Use of Delays

Delays are usually an undesirable effect of the natural world that we try to combat by modifying device processes and circuit configurations. We strive to make circuits faster to solve problems quicker, but the intrinsic delay in circuits hinders this progress and may cause errors in computed outputs, as will be shown later. Although overall delays are an unwanted nuisance, some circuit designers have found various ways to utilize this unwanted characteristic in a myriad of applications. An example of such an application is shown in Figure 4.8—intrinsic inverter delays are used to generate a clock signal.

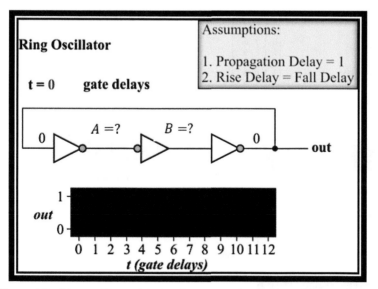

Figure 4.8   Ring oscillator with three inverters and a given starting state at $t = 0$.

Figure 4.8 shows three inverters in a chain with the output of the last inverter fed back to the input of the first. The middle inverter uses a different symbol, but its operation is still the same as the others. (Remember bubble logic from the previous chapter). For the middle inverter, only an asserted LOW signal will cause a HIGH on its output, otherwise, the output remains LOW. For the remaining two symbols, only an asserted HIGH signal will cause a LOW on their outputs, otherwise, their outputs remain HIGH. So all three symbols describe inverters no matter where the bubble goes.

In Figure 4.8, we are given three inverters with an equal propagation delay of 1 unit and the symmetric rise and fall delays. Measuring the output at time $t = 0$, we know that $out = 0$. We currently do not know the intermediate nodes A and B, so they are given the "?" symbol to denote UNKNOWN state. At time $t = 1$, $out = 0$ is propagated to intermediate node A, so node A switches to 1. At time $t = 2$, $A = 1$ is propagated to node B, so node B switches from ? to 0. At time $t = 3$, $B = 0$ is propagated to out, so out switches from 0 to 1. Continuing this analysis through the various time steps will yield that at time $t = 6$, out will switch from 1 back to 0, as shown with the truth table in Figure 4.9. Continuing through the analysis for even more time steps will show that a square wave with a period $T = 6$ units results from the three

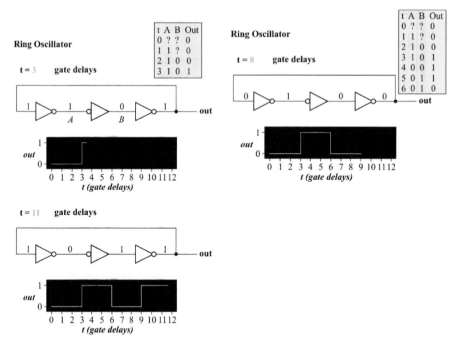

Figure 4.9   Ring oscillator shown at different time steps with truth tables to verify operation for skipped time steps.

inverter ring oscillators. This circuit is an oscillator because it is output *out* bounces between 0 and 1 with a specified period T.

For the general case, the ring oscillator can only be made with an odd number of inverters, n. You can verify this yourself. The period of the resultant clock signal generated from the ring oscillator is given in Equation 4.2. Clocks of different frequencies or periods can be generated through inverter delay using the relationship between the inverter delay t_d and the number of inverters. In our example, three inverters with $t_d = 1$ time unit produced a signal with period $T = 6$ time units and frequency $f = \frac{1}{6}$ (time unit)$^{-1}$. For example if a time unit were 1 ns, then our three-inverter ring oscillator would have a frequency of oscillation as $f \cong$ 166.7 MHz.

$$T = 2 \times n \times t_d \tag{4.2}$$

$$n = 2 \times k + 1 \quad k = 1, 2, 3, \ldots \tag{4.3}$$

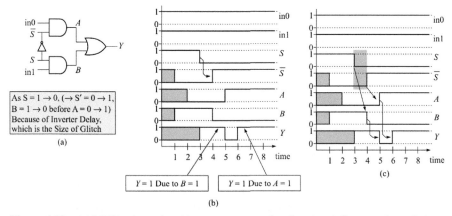

Figure 4.10 (a) MUX schematic with reconvergence showing signal S passes through three gates (inverter, AND, and OR) in one path and two gates (AND and OR) in the other path to realize Y. (b) Timing diagram showing that inverter delay causes a period between time $t = 3$ and time $t = 4$ where S and $\overline{S}$ equal 0. (c)Timing diagram showing that the delay in (b) that caused both S and $\overline{S}$ to equal 0 causes, the output Y to be incorrect between time $t = 5$ and time $t = 6$.

4.5 Glitches and Hazards: Bad Effects Due to Unequal Path Delays of Reconvergent Signal Variables

Without delay, the ring oscillator configuration shown would not work. There always seems to be a good side and a bad side to every story, so this section will introduce the adverse effects of delay through an example of a multiplexer (MUX). The MUX in Figure 4.10a is composed of two AND gates and an OR gate, and an inverter is used to generate a complementary signal. The Boolean expression of the MUX is expressed in Equation 4.4:

$$Y = in0 \cdot \overline{S} + in1 \cdot S \qquad (4.4)$$

The MUX is a circuit that selects one out of multiple data inputs. In this case, we have two data inputs, $in0$ and $in1$. So the output Y is always the value of either $in0$ or $in1$ depending on the value of the select input S. If $S = 1$, then $\overline{S} = 0$ so this forces the relationship, $Y = in1$. If $S = 0$, then $\overline{S} = 1$, so in this case the output will be set as $Y = in0$. In Figure 4.10a, since S is split between two different paths to generate the values at the intermediate nodes A and B, which in turn determine output Y, this circuit configuration is termed a circuit with *reconvergence*. Circuits with reconvergence runs the risk

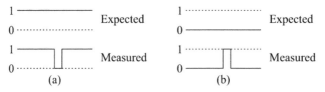

Figure 4.11   (a) Static-1 hazard (expect a 1, but then there is a 0 measured somewhere). (b) Static-0 hazard (expect a 0, but then there is a 1 measured somewhere).

of causing glitches and errors if the delays between both paths are unequal, as shown in Figure 4.10.

In Figure 4.10b,c, the inputs $in0$ and $in1$ are kept at Logic 1 the whole time, so the analysis of this circuit *should* yield that $Y = 1$ despite the value of S. This is not the case, though, as seen in the waveform for Y between time $t = 5$ and time $t = 6$. In this example, all gates have a propagation delay of 1 time unit. At time $t = 3$, S switches from 1 to 0, but it takes an inverter delay of 1 time unit for $\overline{S}$ to switch. During this period, both S and $\overline{S}$ equal 0. Following Figure 4.10c, the difference in delay between the two paths causes the output Y to be incorrect for the same duration as that difference in delay. This circuit is diagnosed to exhibit a *static-1 hazard*, which means the output is expected to remain at Logic 1 but then there is a momentary lapse where the output toggles to 0 and then back to 1. There also exists a *static-0 hazard*, where the output is expected to be 0, but then exhibits a momentary lapse where it toggles to 1 and then back to 0 (see Figure 4.11).

4.5.1 Correction of the static-1 hazard

As explicated above, if timing is not taken care of at the time of design, a digital circuit implementation of a Boolean function may result in some erroneous output. To deal with the issue of timing, we will discuss a process for realizing a Boolean expression for a MUX.

First, we construct a truth table for the MUX, as shown in Figure 4.12a. If $S = 0$, then $Y = in0$. If $S = 1$, then $Y = in1$. The truth table is then converted into a Karnaugh map (K-map) representation, which is discussed in detail in Chapter 6.

In the K-map representation, the inputs are written outside, while the outputs are written in the boxes, as shown in Figure 4.12c. The values for S are on the side, while $in1$ and $in0$ are combined to form a 2-bit number. Make sure you can identify each row of the truth table in the K-Map in Figure 4.12c.

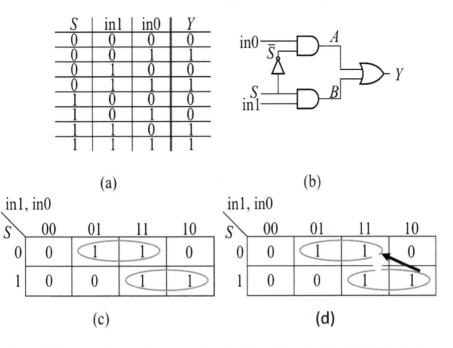

S	in1	in0	Y
0	0	0	0
0	0	1	1
0	1	0	0
0	1	1	1
1	0	0	0
1	0	1	0
1	1	0	1
1	1	1	1

(a)

(b)

(c)

(d)

Figure 4.12 (a) Truth table for a MUX, (b) the MUX schematic, (c) the MUX truth table converted to a K-map representation, and (d) the MUX K-map representation showing the switch from $S = 1$ to $S = 0$.

The K-map is a visual tool you will use later to aid in minimizing Boolean functions. From the K-map, we can identify the group of 1s and circle them, as shown in Figure 4.12c. From the top oval, we can see that the commonality between these two rectangles is that $S = 0$ and $in0 = 1$.[1] From the bottom oval, we can see that the commonality between these two rectangles shows that $S = 1$ and $in1 = 1$. These two ovals correspond to the two AND gates, and both ovals cover all cases where $Y = 1$. This verifies Equation 4.4 as the minimum expression for the MUX.

The reason the K-map is introduced is that it gives insight into the areas our circuits may experience timing errors, as shown in Figure 4.12d. Our circuit in Figure 4.12b, obtained from the K-map analysis, will run into a static-1 hazard while switching from one oval to the other. Although the MUX design introduced covered all cases where $Y = 1$, if the transition from one

---

[1] Note that $in1$ is 0 in the left rectangle but then 1 in the right rectangle, so we do not include it as a common variable between those two rectangles.

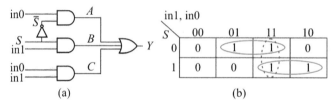

Figure 4.13   (a) Modified MUX schematic with no static-1 hazard and (b) K-map of modified MUX, where the dotted oval introduces an AND gate that takes care of the static-1 hazard when switching from $S = 1$ to $S = 0$.

oval to another is not covered, then we will run into a timing hazard. To correct this error, we introduce another oval, as shown in Figure 4.13b.

In Figure 4.13b, the switching path is covered with another oval. The commonality between the rectangles covered by this dotted oval is $in1 = 1$ and $in0 = 1$. This oval adds an AND gate to the MUX expression, giving us Equation 4.5. Note that Equation 4.5 is not a minimized Boolean expression because of the redundancy presented in the K-map of Figure 4.13b. But minimized expressions do not always yield the best designs when non-ideal effects like timing are considered.

$$Y = in0 \cdot \overline{S} + in1 \cdot S + in0 \cdot in1 \tag{4.5}$$

Finally, it may be noted that if there exist multiple paths between an input (A) and an output (Z) in a large Boolean network, then a dynamic hazard may occur. A dynamic hazard occurs if due to change of logic value in the input (A), the output (Z) undergoes multiple changes of 0 and 1 before it settles to a stable logic level. Note that in Static-0 hazard, the output changes momentarily to 1, resulting in $0 \rightarrow 1 \rightarrow 0$, between two stable states of 0. In case of dynamic hazard, the Boolean network may undergo multiple momentary transitions of 0 and 1 between two different stable output states, $0 \rightarrow 1 \rightarrow 0 \rightarrow 1$. Examples of dynamic hazards occurring in a switching network are illustrated in Figure 4.14. Dynamic hazards are difficult to troubleshoot in a large Boolean network. However, by eliminating static-0 and static-1 hazards in a Boolean network, the risk of dynamic hazards can be potentially reduced.

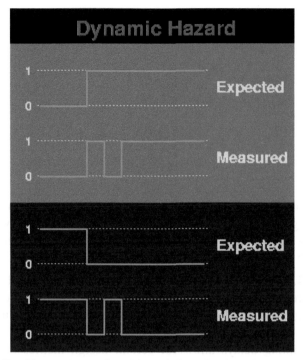

Figure 4.14 Two illustrated examples of dynamic hazards where the output makes multiple transitions before settling to its correct stable value.

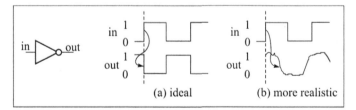

Figure 4.15 Variant models showing that signals do not change instantaneously and Logic 0 and Logic 1 values do not mean signals behave in a square-wave fashion.

4.6 Conclusion and Key Points

To wrap up this chapter, we will recapitulate and link what we have learned with the following points:

1. In real life, logic gates behave in a nonideal fashion. They have propagation delays, their signals have finite rise and fall times, and

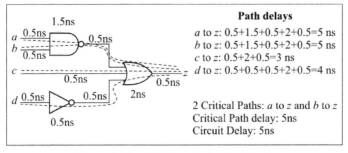

Figure 4.16   Critical paths for the circuit.

their output signals do not necessarily look like a square wave. These properties give rise to issues like static hazards. However, clever circuit design can also leverage the property of time delay to our benefit, such as building simple oscillators based on gate delays.

In Figure 4.15, ideal input-output signalsare juxtaposed with realistic waveforms to remind you that signals do not change instantaneously and Logic 0 and Logic 1 values do not necessarily mean signals behave in a pretty square-wave fashion.

The finite timing delays allow for certain design metrics that digital designers adhere to:

- *Path delays* can be calculated for any path from input to output (sum up all delays while traversing the path of interest).
- The *critical path* is the path that exhibits the longest delay, and *critical delay*or *critical path delay* is the delay associated with the critical path.
- Figure 4.16 shows two critical paths exist for the circuit, the paths Az and Bz, since they are the paths with the largest delay. Path delay of cz is 3 ns and of dz is 4 ns. The *circuit delay* takes on the value of the critical path delay.

Problems on Timing Diagrams

1. For the given digital circuit, identify the correct output for the given input waveforms. Assume that all gates have zero delay.

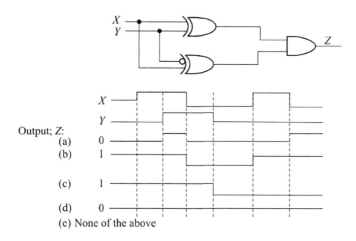

Output; Z:
    (a)   0
    (b)   1

    (c)   1

    (d)   0
    (e) None of the above

2. For the given digital circuit, draw the output waveform for the given input waveforms. Assume that all gates have zero delay. Note that b $(= 0)$ does not change with time.

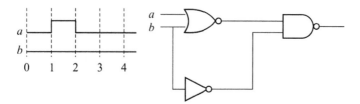

3. For the same circuit of problem #2 with the given input waveforms, draw the output waveform. Assume that the NOR and NAND gates have two units of delay and the inverter has one unit of delay.

4. Given the following gated network, write an expression for Z and simplify. Assume that each gate has one unit of delay. (Delay time = 1 on the timing diagram.) Draw the output waveform (Z) for the given input values of X and Y.

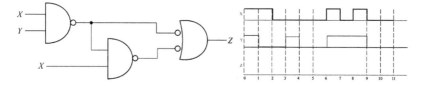

5. Draw the output waveform (Z) for the given values of X and Y. Assume each gate has zero delay.

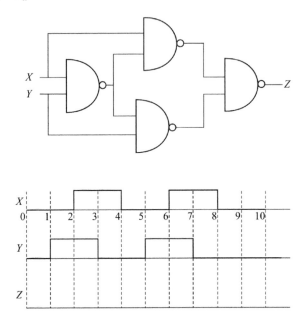

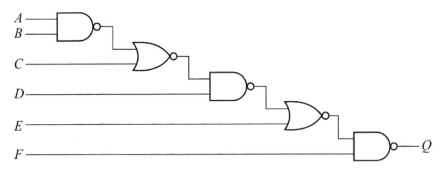

6. Given the circuit below and the following gate propagation delays:

$$t_{pHL}^{NAND} = 0.5 \text{ ns}; \ t_{pHL}^{NOR} = 1.0 \text{ ns}$$

$$t_{pLH}^{NAND} = 0.75 \text{ ns}; \ t_{pLH}^{NOR} = 1.25 \text{ ns}$$

compute $t_{pHL}^{A \rightarrow Q}$ and $t_{pLH}^{A \rightarrow Q}$ under the following input combination:
$B = 1, C = 0, D = 1, E = 0, F = 1$.

7. Given the circuit below with the gate delays listed on the figure, draw the output waveform and calculate the input-to-output delay for the given inputs.

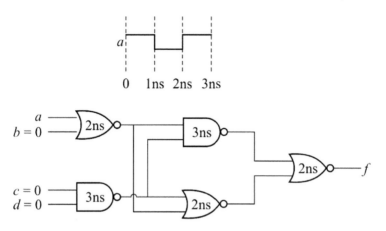

8. For the circuit given in problem #7, find an input combination/transition that causes the critical delay (i.e., the longest possible input-to-output delay).

9. Given the circuit below with inverter delays $t_{pHL} = t_{pLH} = 1$ ns and all other delays $t_{pHL} = t_{pLH} = 2$ ns:

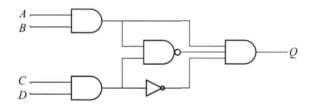

10. Assume that A has a rising transition (i.e., $A = 0 \rightarrow 1$). What fixed input values are necessary on B, C, and D such that there is a transition at the output Q? What is the propagation delay $t_{pLH}{}^{A \rightarrow Q}$?

11. What are $t_{pHL}{}^{C \rightarrow Q}$ and $t_{pLH}{}^{C \rightarrow Q}$, given inputs $A = B = D = 1$? Show a timing diagram for each case.

12. Is this circuit free of static hazards? Explain.

13. The following waveforms show the simulation trace of a two-input NAND. Find the rise and fall delays of the gate.

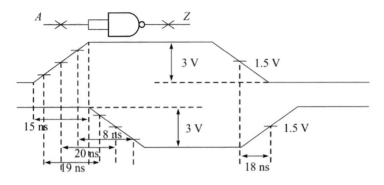

The delay, t_{pHL}, of the gate is:
A. 15 ns
B. 18 ns
C. 19 ns
D. 20 ns
E. None of the above

Combinational Logic Design Techniques: Part I

Given a problem statement in English, what tools are necessary to transform this problem statement into a combinational logic solution? This chapter will strive to bridge the gap between a problem and the solution by showing through examples the combinational logic design process. The first example, the stairwell lamp problem, will be dealt with in much detail, while the next examples will apply the steps introduced in a speedier fashion. A counterexample is then presented to show that although the design process is laid out, intuitive understanding of the problem statement is necessary to conserve time and design effort during a design process.

Terms introduced in this chapter: truth table, canonical SOP, minterms, minterm expansion, true, complement, BCD, and logic completeness

Competency Objectives: At the end of this chapter, you will be able to:

1. Design logic circuits that exhibit desired behaviors in the real world.
2. Utilize truth tables to document all possibilities of a logic function.
3. Describe logic functions as canonical sum of products with minterms.
4. Learn how, at times, we can adopt intelligent alternatives to truth table analysis that may enable designing logic circuit faster.

5.1 Designing a Digital System from a Problem Statement

5.1.1 Stairwell lamp problem

Problem: In a three-story building, there is a lamp to illuminate a stairwell. The lamp can be independently turned ON and OFF from each floor by flipping an electrical switch on that floor. Design the logic circuit for the problem.

Solution:

Step 1: Assign Boolean variables and define their values with respect to the problem.

In this problem, the three electrical switches can be in one of two positions, UP position or DOWN position. Notice that we do not define the positions as ON or OFF, since we have the requirement that each switch be able to turn the lamp ON or OFF, independently from the other switches. So, for example, flipping the switch on the first floor to the UP position may either turn the lamp ON (if the lamp were previously OFF) or turn the lamp OFF (if the lamp were previously ON).

Our first set of Boolean variables, A, B, and C, will keep track of the switch positions for the first, second, and third floors, respectively. Only one more Boolean variable, L, is needed to denote the status of the lamp. At this point, each of the three inputs along with the output are assigned a unique variable.

We proceed further to define the meaning of the Boolean states. Let each switch, $X \in \{A, B, C\}$, have two positions, UP and DOWN, and let the lamp, L, have two states, ON and OFF. We arbitrarily choose $X = 1$ to denote a switch is in the UP position and $X = 0$ to denote a switch is in the DOWN position.[1] Likewise, we choose $L = 0$ to denote that the lamp is OFF and $L = 1$ to denote that the lamp is ON.

Step 2: Write the truth table.

The *truth table* is a tool that shows the input–output relationships between Boolean variables. A complete truth table will list all possible input–output combinations. For our stairwell example, we have three input Boolean variables (A, B, and C) and one output variable (L). We can write a Boolean

---

[1]Consistency is the name of the game. $X = 0$ can be chosen to represent UP, and $X = 1$ can be chosen to represent DOWN.

function *L* with input parameters *A*, *B*, and *CL(A,B,C)*. Each input can take at most two values, 0 or 1, so three inputs will have $2^3 = 8$ rows in a complete truth table, corresponding to the total number of distinct input combinations.

In Figure 5.1, each row of the truth table signifies an input combination and its resulting output. To make sure that we do not miss any possible input combination, the rows are numbered 0, 1, 2, 3, 4, 5, 6, and 7, corresponding to the decimal number that can be obtained if *ABC* were seen as a 3-bit binary number with *A* as the *most significant bit* (MSB) and *C* as the *least significant bit* (LSB).

To generate the truth table, we establish a starting point (or reference position) and then determine the other cases on the basis of the starting point. A reference position is needed in this problem because the problem statement does not give any other relationship between the output and inputs except the ability to turn the lamp ON or OFF from any floor. The starting point chosen here is the 000 case when all three switches are DOWN; we define the lamp as being in an OFF state.[2] In American convention, electrical switches in the DOWN position usually turn a lamp OFF, so this is partly why we chose this convention. The other reason stems from previous logical habits where 0 usually signifies OFF, so if all switches are 000 then we automatically think OFF. Remember, though, that the definition of 0 for the switches means DOWN and not OFF.

At this point, we will define four states: zero-UP, one-UP, two-UP, and three-UP. These states correspond to the number of switches that are in the UP position; for example, zero-UP has no switch in the UP position, so it corresponds to the 000 case, and three-UP has three switches in the UP position, so it corresponds to the 111 case. So continuing the analysis from the zero-UP state, we see that if any one of the switches is flipped to an UP position, that is, if we change states to a one-UP state (the 001, 010, or 100 case), then the lamp will turn ON. From the one-UP state we can switch the lamp OFF by either going back to the zero-UP state or flipping another switch to the UP position, which makes us end up in the two-UP state (the 011, 101, or 110 case). From the two-UP state we may turn the lamp back ON by either going back to the one-UP state (put an UP switch in the DOWN position) or going forward to the three-UP state (put a DOWN switch in the UP position). Arrows are drawn in Figure 5.1 to show the transition from one

---

[2]We could have chosen the 000 state to make the lamp be in the ON state. The problem statement did not limit our choices here. Again, be consistent after choosing a convention and you will be fine.

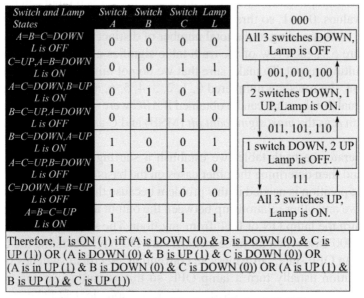

Figure shows the result of the truth table for *L* (*A*,*B*,*C*).

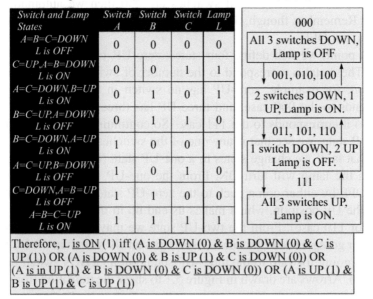

Figure 5.1 (Left) Truth table for the stairwell problem. (Right) State diagram showing transition from one state to another. (Bottom) Verbally identified situations that constitute the lamp being ON.

state to another, and this state transition allows us to determine our Boolean values for the output L in the truth table.

Step 3: Write the canonical sum of products (standard SOP) for the output Boolean variable(s).

From the truth table in Figure 5.1, the output L is equal to 1 for either one of the four following switch combinations (001, 010, 100, and 111). Only one of these four combinations needs to be true for the lamp to be ON, so we have an expression joined by OR statements:

$$L = 1 \text{ iff}^3 \ (A = B = 0, C = 1) \text{ OR } (A = C = 0, \ B = 1)$$
$$\text{OR } (A = 1, \ B = C = 0) \text{ OR } (A = B = C = 1).$$
$$L(A, B, C) = \overline{A}\overline{B}C + \overline{A}B\overline{C} + A\overline{B}\overline{C} + ABC \tag{5.1}$$

The expression in Equation 5.1 is the canonical *sum of products* (SOP) or the standard SOP for the Boolean variable L. The expression is called an SOP because we have product terms (AND operations) that are summed (OR operations) together. In the *canonical* form, each product term, for example, $\overline{A}B\overline{C}$, has in its expression all input variables in either their *true* (e.g., A) or *complement* (e.g., $\overline{A}$) form. The canonical form may not constitute a minimized Boolean form.

Another name for inputs in this case is *literal*, which applies to the input, whether it exists in the true or in the complement form. The output representation in Equation 5.1 can be written as a sum of *minterms*, and in this form, it is called a *minterm expansion*, as shown in Equation 5.2a–c.

$$L(A, B, C) = \sum (1, 2, 4, 7) \tag{5.2}$$
$$L(A, B, C) = \sum m(1, 2, 4, 7) \tag{5.3}$$
$$L(A, B, C) = m_1 + m_2 + m_4 + m_7 \tag{5.4}$$

Minterms correspond to the decimal representation of the binary number constituted by the input variables in the truth table. From the truth table in

---

^{3}iff: if and only if.

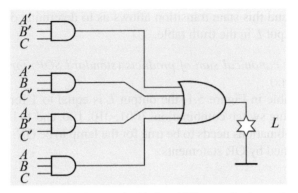

Figure 5.2   Canonical SOP implemented with four AND gates for the product terms and one OR gate.

Figure 5.1, minterm #0, or m_0, corresponds to the 000 row; minterm #1, or m_1, corresponds to the 001 row; and so on. The minterm expansion form directly expresses the canonical SOP of the logic expression.

Step 4: Implement the canonical SOP expression in logic circuit.
Take the expression in Equation 5.1 and directly draw it with actual gates, as shown in Figure 5.2.

The direct expression involves four three-input AND gates and one four-input OR gate to implement the function L. To determine if this realization is a good design, three metrics (*gate cost*, *literal cost*, and *transistor count*) are used to evaluate the cost of implementation:

- Gate cost: Count the number of gates, noting the number of inputs for each gate. In Figure 5.2, the implementation has 4 × (3-input AND) and 1 × (4-input OR), so 5 gates in total. *Note: This method of comparison is usually not advisable because direct information is not present.* For example, how does the size of the three-input AND gate compare to that of a two-input AND gate or a four-input OR gate? Gates usually differ in size, depending on the number of inputs and the type of gate; therefore, a direct gate count does not relay that much information.

- Literal cost: Count the number of literals present. The four three-input AND gates have three literals each at their inputs, so the literal count for the current design computes to 3 × 4 = 12 literals. *Note: You cannot reuse literals, for example, the input C is counted as a literal for the top AND gate and it is also counted as a literal for the bottom AND gate.* This method of accounting helps embed wire costs into our literal cost evaluation metric.
- Transistor count: Determine the total number of transistors necessary to implement the entire design. In standard complementary metal-oxide semiconductor (CMOS) technology, logic gates generally have inverted outputs. To implement an N-input NAND or NOR gate, $2N$ transistors are required. To implement an AND or an OR gate, an additional NOT gate using two transistors is needed. Hence, NAND and NOR gates (instead of AND and OR gates) are more commonly used in a CMOS implementation of Boolean functions. The transistor rule of thumb allows us to see that a three-input AND is implemented with $(2 \times 3) + 2 = 8$ transistors, and a four-input OR is implemented with $(4 \times 2) + 2 = 10$ transistors. We have four three-input AND gates, which entails $4 \times 8 = 32$ transistors, and one four-input OR gate, which has 10 transistors, which sums up to a total of 42 transistors for the canonical SOP implementation.

Note: Remember, besides AND and OR gates, we have other gates at our disposal, notably the NAND, NOR, XOR, and XNOR gates. Recapping from the Boolean algebra chapter, Figure 5.3 provides a list of the other gates and their Boolean equations.

Remember, CMOS NAND and NOR gates are duals of one another. In NAND, the P-type metal-oxide semiconductors (PMOSs) are in parallel and the N-type metal-oxide semiconductors (NMOSs) in series, while in NOR the PMOSs are in series and the NMOSs in parallel. PMOSs are good at passing 1, while NMOSs are good at passing 0. The best way to realize AND and OR gates is to use a NOT gate at the output of the NAND and NOR gates. This is the reason AND and OR gates have two more transistors than the NAND and NOR gates.

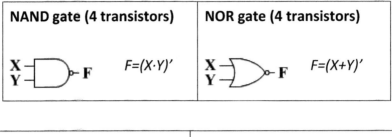

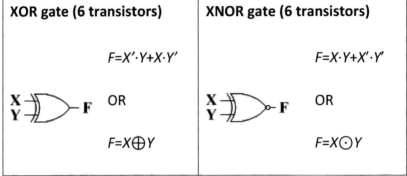

Figure 5.3   NAND, NOR, XOR and XNOR gate representations with their equations.

Step 5: Minimize the canonical SOP.
Use Boolean algebra to simplify the following canonical expression:

$$
\begin{aligned}
L\ (A,\ B,\ C) &= AB'C' + A'BC' + A'B'C + ABC \\
&= (AB'+A'B).C' + (A'B' + AB).C \quad \text{(Law of Distribution)} \\
&= (A \oplus B).C' + (A \oplus B)'.C \quad \text{(Def. of Ex-Or Equivalence)} \\
&= (A \oplus B) \oplus C = A \oplus B \oplus C.
\end{aligned}
$$

Gate cost: 2 × 2-input XOR

Literal cost: 3

Transistor count: 6 × 2=12

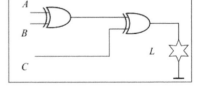

Comparison with the direct canonical SOP implementation:
Gate cost: $(5 - 2)/5 \times 100 = 60\%$ reduction of gate count

Literal cost: $(12 - 3)/12 \times 100 = 75\%$ reduction of literal cost

Transistor count: $(42 - 12)/42 \times 100 = 75\%$ reduction of transistor count

Minimization is important because minimized logic implementations use fewer resources than the direct canonical SOP implementation. For example, consider how many minterms appear in the expression when instead of 3 floors we have 11 floors? As shown in this example, 3 floors yield 4 terms, but 11 floors would yield 1024 terms [we need to introduce 11 different variable corresponding to each of the 11 floors, so there will be $2^{11} = 2048$ possible combinations of inputs, half of the combinations ($\frac{2048}{2} = 1024$) will correspond to an ON state of the lamp]. The output expression grows exponentially, so conservation of resources is necessary.

A general solution then for an *n*-story building is as follows:

$$L(S_0, S_1, S_2, \Box, S_{n-1}) = S_0 \oplus S_1 \oplus S_2 \oplus \Box \oplus S_{n-1}$$

Representations: To recap, there exist multiple equivalent representations (truth table, equations, and circuits) of the solution to a given problem statement. We can convert between the representations in either ways.

In the next section, a binary-coded decimal (BCD) to seven-segment converter example is discussed.

5.1.2 BCD to seven-segment converter

BCD is a number representation in which each digit is represented by its own binary sequence. For example, the number 159 is represented as 0001 0101 1001: 0001 is the number 1 in binary, 0101 is the number 5 in binary, and 1001 is the number 9 in binary. The BCD representation is usually converted to a seven-segment display for visual representation in electronics, for example, microwave ovens, digital watches, etc.

The example in Figure 5.4 shows a BCD number "*wxyz*" as an input to a converter that produces the output for the seven-segment display "*abcdefg*." Each letter can be either 0 (OFF) or 1 (ON). Two examples are shown for the numbers 3 and 5, where the values for "*abcdefg*" correspond to 1111001 and 1011011. Figure 5.5 shows how we can divide an oscillator's output signal to feed a 7-segment display.

The internal clock oscillates at 2^{16} times every second. The clock is then consecutively divided by 2 multiple times until it oscillates 1 time every second (1 Hz). The clock is then divided by 60 to obtain an oscillation frequency of 1 change every minute. This oscillation frequency can be used to determine the minute digit in a display. The frequency of 1 change every minute is divided by 10 to obtain 1 change every 10 min, and this can be used to determine the tens of minute digit on the clock display. The frequency of 1 change every 10 min is then divided by 6 to obtain 1 change every 60 min or 1 change every 1 h. This resultant frequency is used to determine the hour digit, and dividing this frequency by 10 will yield 1 change every 10 h. A frequency of 1 change every 10 h is used to obtain the tens of hour digit on the digital display. There is indeed a need for a representation such as BCD, and BCD helps transform digital 1s and 0s to something human eye can make sense of. Figure 5.6 shows a truth table used to determine the seven-segment display from the BCD number.

From Figure 5.6, ten digits, that is, 0 through 9, can be represented on the seven-segment display. The minimum number of bits in binary required to represent the largest decimal digit 9 is 4 bits; 111 is the highest 3-bit number, which is 7. With a 4-bit number, we can represent decimal numbers 0 (0000) through 15 (1111). In our BCD expressions, we only represent digits, so the decimal number 10 in BCD would be 0001 0000, constituting the digits 1 and 0. The remaining 4-bit numbers from (1010 to 1111) has no corresponding single digit representation in decimal. Hence we choose to set to keep all the LED of the seven-segment display OFF for all these unused bits. This is purely due to the fact that 4 bits contain more representations than is necessary to code 10 digits (0 through 9).

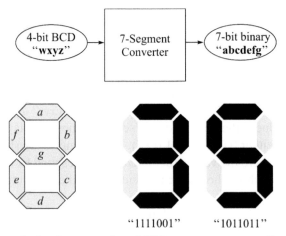

Figure 5.4   Example showing conversion of BCD to a seven-segment display, defining the bit variables for the display.

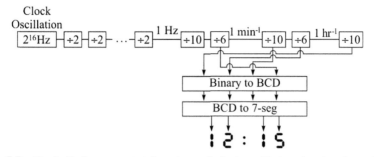

Figure 5.5 Clock display generated from internal clock oscillation showing the path from the internal clock to the digital out.

In the output columns ("abcdefg"), rows 0 through 9 in the truth table determine the segments that turn ON when the BCD number is "*wxyz.*"For example, "*wxyz*" = 1000 represents the decimal number 8, so the seven-segment display should have all segments ON, hence "*abcdefg*" = 1111111. From the truth table representation, the SOP representation can be written as a sum of minterms, as shown with examples in Figure 5.6, for the output variables, *a*, *b*, *c*, and *d*.

digit	wxyz	abcdefg
0	0000	1111110
1	0001	0110000
2	0010	1101101
3	0011	1111001
4	0100	0110011
5	0101	1011011
6	0110	1011111
7	0111	1110000
8	1000	1111111
9	1001	1111011
A	Invalid	0000000
...	...	...
F	Invalid	0000000

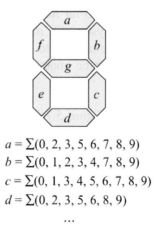

$a = \Sigma(0, 2, 3, 5, 6, 7, 8, 9)$
$b = \Sigma(0, 1, 2, 3, 4, 7, 8, 9)$
$c = \Sigma(0, 1, 3, 4, 5, 6, 7, 8, 9)$
$d = \Sigma(0, 2, 3, 5, 6, 8, 9)$
...

Figure 5.6 Truth table and equations used to determine the Boolean values for the seven segments.

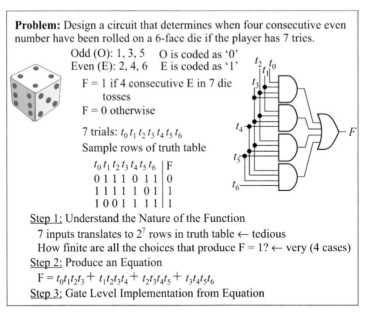

Problem: Design a circuit that determines when four consecutive even number have been rolled on a 6-face die if the player has 7 tries.

Odd (O): 1, 3, 5   O is coded as '0'
Even (E): 2, 4, 6   E is coded as '1'

F = 1 if 4 consecutive E in 7 die
tosses
F = 0 otherwise

7 trials: $t_0 \, t_1 \, t_2 \, t_3 \, t_4 \, t_5 \, t_6$
Sample rows of truth table

$t_0 \, t_1 \, t_2 \, t_3 \, t_4 \, t_5 \, t_6$	F
0 1 1 1 0 1 1	0
1 1 1 1 1 0 1	1
1 0 0 1 1 1 1	1

Step 1: Understand the Nature of the Function
   7 inputs translates to 2^7 rows in truth table ← tedious
   How finite are all the choices that produce F = 1? ← very (4 cases)
Step 2: Produce an Equation
   $F = t_0 t_1 t_2 t_3 + t_1 t_2 t_3 t_4 + t_2 t_3 t_4 t_5 + t_3 t_4 t_5 t_6$
Step 3: Gate Level Implementation from Equation

Figure 5.7   Die pattern detector intuitive implementation.

5.1.3 Event detector

The steps shown so far have given a systematic way to solve logic circuit problems. The systematic method ensures that all possible input variable combinations are taken into account. But sometimes following a problem-solving recipe can be unnecessarily tedious, as shown in the next example (Figure 5.7).

   The goal in Figure 5.7 is to design a logic circuit that can detect four consecutive even numbers in seven die tosses. The first step is to translate this problem into digital computing language. Each die toss or roll constitutes one trial with an outcome that will be stored in a variable. Seven trials are stored in the variables t_0 through t_6. The goal is to provide a YES or a NO if four consecutive even numbers are detected. If yes, the output variable F will be set as 1, otherwise 0.
   A direct implementation of this in truth table format would require 128 rows, so it is definitely impractical to write out by hand. The best way to solve this problem would be to write an equation based on the only possible times F returns YES. Four consecutive even numbers are detected if consequtively t_0, t_1, t_2, and t_3 are 1 or if t_1, t_2, t_3, and t_4 are 1 and so on. This allows us to write

an equation for the output, as shown in Figure 5.7. Using the output equation F, product terms are implemented with AND gates and the sum terms with the OR gate.

5.2 Conclusion and Key Points

The procedure outlined in this chapter can also be summarized in a two-step process. The first step is to describe the function by either creating a truth table or directly implementing equations that portray the desired behavior of each output (depending on the nature of the given problem). The second step is to obtain an equation by directly writing the sum of the minterms, and then implement the gate-based circuit.

- A complete truth table lists all possible input–output combinations and can become very large when considering more than three variables.
- Boolean functions can be represented as canonical SOP and can be implemented as such. Canonical SOP implementations can end up using more gates or hardware and as such, it is often better to minimize canonical SOP expressions using Boolean algebra or some other means to obtain an implementation that utilizes fewer hardware components.
- Sometimes writing out truth tables to derive equations can be unnecessarily cumbersome and tedious. During those times, an alternative approach that takes into consideration insights gleaned from the problem description can enable an equation representation without identifying minterms in a truth table.

Problems on Combinational Circuits: Part I

1. The following gated networks show pairs of logic circuits: (A1, A2), (B1, B2) and (C1, C2). In which circuit pairs do circuit 1 and circuit 2 generate the same logic function?

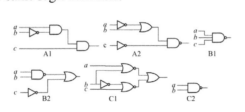

A. (A1, A2), (B1, B2)   B. (B1, B2), (C1, C2)   C. (A1, A2), (C1, C2)
D. All three   E. None of the above

2. Consider the following logical expressions:

    (i) $z(a, b, c, d) = (bd + \bar{a})\bar{c}$
    (ii) $z(a, b, c, d) = [(\overline{bd})a + c]$
    (iii) $z(a, b, c, d) = [(ab + c)d + \bar{a}]\bar{c}$

Which of the above expressions is equivalent to the output of the gate circuit given?

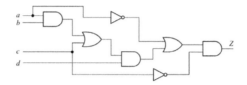

A. Only expression (i)   B. Only expression (iii)   C. All but expression (i)
D. All but expression (iii)   E. All three expressions

3. If m_i is the ith minterm and M_i is the ith maxterm, then $\overline{m_i} + \overline{M_i}$ equals:

A. 0   B. m_i C. M_i D. 1   E. None of the above

4. Design a circuit for a traffic light that takes in the inputs from detectors pointing in the east, west, north, and south directions. The detectors indicate if a car is coming in their direction. The circuit will indicate that the light should turn green if there are cars coming from the east or west directions but not the north or south directions. You can represent the detector outputs with the letters E, W, N, and S, respectively, and the circuit output with the letter G.

5. A TV show has three contestants who are trying to push their buzzers to get a chance to answer the host's questions. Each contestant has a buzzer and an indicator light in front, which turns on if the contestant is the only one to push the buzzer. If two or more contestants push their buzzers at the same time, none of the lights turn on. Design the detector circuits for each contestant's buzzer and indicator light.

6. Determine the minimum number of transistors required to implement the Boolean expression $f(a, b, c) = ab + (a + c')$.

7. Determine the minimum number of transistors required to implement the Boolean expression $f(a, b, c) = ab'c + a'bc$.

8. Using only NAND gates of arbitrary inputs, design a BCD to seven-segment decoder where the BCD inputs are designated by w, x, y, z with w being the MSB. The seven segments of a display device are connected

to the outputs of the decoder. Write expressions for each decoder output. You need not minimize the Boolean expressions for the output.

9. Using NAND gates only, design a circuit that determines whether "three consecutive 5's" occur when a gambler is allowed to make 6 throws consequtively of an unbiased 6-faced die.

   a) Calculate the probability of successful outcome of three consecutive 5's occurring in a trial of 6 throws.
   b) Recompute the probability of success if the outcome is defined by "*exactly* three consecutive 5's".
   c) If you are a hardware designer for a casino owner, what will you recommend to your boss?

   Note that the probability of success is lower with "*exactly* three consecutive 6's". But there are tradeoffs in terms of hardware cost and gambler's psychology. Provide your justification with quantitative data for the above two cases.

10. Using a minimum number of NAND gates only, design a circuit that determines if 4 consecutive black marbles are followed by a white marble if a blindfolded person performs 7 draws consecutively from an urn. Compare the hardware cost if you are asked to redesign your circuit by using minimum number of NAND gates only.

11. Using only NOR gates, design a BCD-to-BCD converter that uses BCD Gray Code as input bits and generates codes as 8421 BCD. Apply DeMorgan's Laws to transform NOT, AND and OR gates into NOR gates. Find the literal, gate and transistor costs of your implementation.

12. Using only NAND gates, design a BCD-to-BCD converter that uses BCD Gray Code as input bits and generates codes as 84-2-1 BCD. Apply DeMorgan's Laws to transform NOT, AND and OR gates into NOR gates. Find the literal, gate and transistor costs of your implementation.

13. An implication gate (IMP gate) has 2 inputs (X and Y) and an output (Z_{IMP}) given by $Z_{IMP} = \overline{X} + Y$. Prove that the IMP gate is functionally complete by showing that by using at most two IMP gate a 2-input NAND function can be obtained where each input of the NAND gate is in True form only.

14. Show the truth table of a 3-input CRAZY (A, B, C) gate which gives an output of 1 iff A = B, and C = 1. Write the canonical SOP and POS expressions of the CRAZY gate at first and then minimize the SOP and POS expressions. Compare the literal and transistor count costs for minimal SOP and POS expressions.

15. A switching network uses *two* control inputs ($C1$, $C2$) to select an output functionof two data inputs ($X1$, $X2$). The network performs an Implication function, $X1 \subset X2$ when $C1 = C2 = 0$, an equivalence function, $X1 \odot X2$ when $C1 = C2 = 1$, an Inhibition function, $X1/X2$ when $C1 = \overline{C2} = 0$, and an exclusive OR, $X1 \oplus X2$ when $\overline{C1} = C2 = 0$. Draw the switching network with NAND gates only. Estimate the total number of transistors in the logic gates needed to implement the above switching functions.

16. Using only 2-input NAND gates, design a BCD-to-seven-segment display decoder according to the following table. Note that the seven outputs of the seven segment decoder are *a, b, c, d, e, f,* and *g,* which are functions of inputs *w, x, y,* and *z.*

digit	wxyz	abcdefg
0	0000	1111110
1	0001	0110000
2	0010	1101101
3	0011	1111001
4	0100	0110011
5	0101	1011011
6	0110	1011111
7	0111	1110000
8	1000	1111111
9	1001	1111011
A	Invalid	0000000
...	...	...
F	Invalid	0000000

$a = \Sigma(0, 2, 3, 5, 6, 7, 8, 9)$
$b = \Sigma(0, 1, 2, 3, 4, 7, 8, 9)$
$c = \Sigma(0, 1, 3, 4, 5, 6, 7, 8, 9)$
$d = \Sigma(0, 2, 3, 5, 6, 8, 9)$
...

17. Implement the following set of equations using only 2-input NAND gates. Draw the truth table for the problem like in the previous example for BCD-to-seven-segment decoder. Compare the number of gates and number of literals of the two implementations of the decoder. Note that the output *a* depends on another output *c*, which depends on the output, *e*, which depends on *d* and *b*. Since *d* depends on *a*, there is a cycle: *adeca.*

$$g = x_0\bar{x}_3 + a$$

Combinational Logic Design Techniques: Part II

This chapter will continue to reinforce the combinational logic design method/technique introduced in the previous chapter. It will present an alternative (the product of sums, POS) design approach to the sum of products (SOP) design approach. The stairwell lamp problem and the majority gate example of this chapter are implemented in a hardware description language called Verilog. Finally, the Verilog description of a ripple carry adder is presented.

Terms introduced in this chapter: canonical POS, maxterm, maxterm expansion, HDL, full adder, hierarchical design, ripple carry adder

Competency Objectives: At the end of this chapter, you will be able to:

1. Identify Boolean expressions for majority gates in either SOP or POS implementations.
2. Define self-duality and identify Boolean functions that are self-dual.
3. Describe why different implementations, e.g., POS and SOP implementations, are used in logic design.
4. Describe logic functions (majority, full adder, and ripple carry adder) using hardware description language.

6.1 Majority Gate Design

Problem: An n-input majority gate has an output $Z = 1$ iff at least $\lfloor n/2 \rfloor + 1$ inputs are 1. Otherwise, $Z = 0$. For a 3-input majority gate (A, B, and C), at least 2 inputs must be 1 in order for the output $Z(A,B,C) = 1$.

Solution:

From the stairwell lamp problem, the design steps are:

1. Assign Boolean variables and define their values with respect to the problem.
2. Write the truth table.
3. Write the canonical sum of products (SOP) for the output Boolean variables.
4. Directly implement the canonical SOP expression.
5. Minimize the canonical SOP.

In this majority gate example, steps 1 and 2 will be presented as in the stairwell lamp problem, but alternatives will be presented for steps 3, 4, and 5. Step 1 is already done in the problem description: the inputs are A, B, and C, and the output is Z. To complete step 2, we note that, whenever two or more inputs are 1, the output Z is 1. Figure 6.1 shows the corresponding truth table.

In Figure 6.1, the truth table of the majority gate is shown with A, B, and C as inputs and Z as the output. To the left of the truth table, the *minterms* and *maxterms* are explicitly expressed. Minterms, previously defined, are terms that identify the input combination based on the decimal value of the combination's binary representation. Maxterms, on the other hand, also correspond to the decimal value of the binary representation; but instead of the inputs combined with AND gates, the inputs are combined with OR gates. Minterms depend on the logical rule that the combination of input A, input B, and input C will make the output Z true. Maxterms, on the other hand, depend on the combination of input A, input B, and input C that will make the output Z false. We shall discuss further on Maxterms in the later part of this chapter, i.e., in Section 3.1.2.

To compare design with minterms and maxterms, two different approaches will be used to implement the majority gate. The first approach, numbered {3a, 4a, and 5a}, will follow the previously outlined steps to obtain

Minterm and Maxterm	A	B	C	Z
m0 = A'B'C' M0 = A+B+C	0	0	0	0
m1 = A'B'C M1 = A+B+C'	0	0	1	0
m2 = A'BC' M2 = A+B'+C	0	1	0	0
M3 = A'BC M3 = A+B'+C'	0	1	1	1
M4 = AB'C' M4 = A'+B+C	1	0	0	0
m5 = AB'C M5 = A'+B+C'	1	0	1	1
m6 = ABC' M6 = A'+B'+C	1	1	0	1
m7 = ABC M7 = A'+B'+C'	1	1	1	1

Figure 6.1 Majority gate truth table showing all minterms and maxterms.

the canonical SOP, while the second approach, numbered {3b, 4b, and 5b}, will obtain the canonical POS expressions.

6.1.1 SOP implementation

After obtaining the truth table, step 3a involves finding the SOP expression. From the truth table, Z is ON iff ($A = 0$ and $B = 1$ and $C = 1$) OR ($A = 1$ and $B = 0$ and $C = 1$) OR ($A = 1$ and $B = 1$ and $C = 0$) OR ($A = 1$ and $B = 1$ and $C = 1$)

$$Z\left(A, B, C\right) = \overline{A}BC + A\overline{B}C + AB\overline{C} + ABC = \Sigma m(3, 5, 6, 7) \quad (6.1)$$

Equation 6.1 captures the canonical SOP expression for Z, also written in the minterm expansion form for further clarity.

Step 4a implements the SOP expression with AND and OR gates, as shown below:

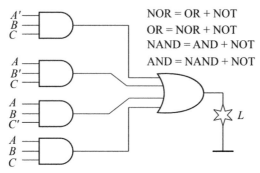

NOR = OR + NOT
OR = NOR + NOT
NAND = AND + NOT
AND = NAND + NOT

Cost of implementation:

Gate cost: Four three-input AND gates + One four-input OR gate = 5 gates.

Literal cost: $4 \times 3 = 12$ literals.

Transistor count: Using the complementary metal-oxide semiconductor (CMOS) rule, that is, an N-input NAND or NOR gate requires $2N$ transistors and an N-input AND (NAND + NOT) or OR (NOR + NOT) gate requires $2N + 2$ transistors, we can obtain the transistor count for implementation: $4[3 \times 2 (\text{NAND}) + 2 (\text{NOT})] + 1[4 \times 2 (\text{NOR}) + 2(\text{NOT})] = 42$ transistors.

The expression can be reduced further using Boolean algebra to implement step 5a.

$Z (A, B, C) = A'BC + AB'C + ABC' + ABC$

$= A'BC + ABC + AB'C + ABC + ABC' + ABC$ (Idempotent Laws)
$= (A+A')BC + A(B+B')C + AB(C+C')$ (Distributive Laws)
$= 1.BC + A.1.C + AB.1$ (Laws of Complementarity)
$= BC + CA + AB$ (since, $Y.1 = Y$; Commutative Laws)

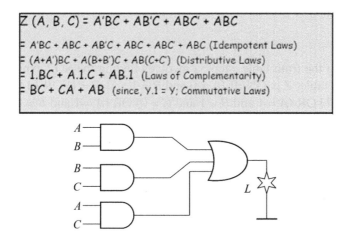

Cost of implementation:

Gate cost: 4 (savings of 20% compared to the canonical implementation).

Literal cost: $3 \times 2 = 6$ (savings of 60% compared to the canonical implementation).

Transistor count: $3[2 \times 2$ (NAND) $+ 2$ (NOT)$] + 1[3 \times 2$ (NOR) $+ 2$(NOT)$]$ $= 26$ (savings of 16% compared to the canonical implementation).

6.1.2 POS implementation

Now we will show the second approach, steps 3b, 4b, and 5b, to implementing the majority gate. After obtaining the truth table in Figure 6.1, we focus on the entries that make the output $Z = 0$, and this will enable us to complete step 3b, finding the canonical POS expression.

Z is OFF, that is, Z' is ON, iff ($A = 0$ and $B = 0$ and $C = 0$) OR ($A = 0$ and $B = 0$ and $C = 1$) OR ($A = 0$ and $B = 1$ and $C = 0$) OR ($A = 1$ and $B = 0$ and $C = 0$).

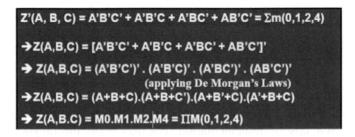

The derived expression is the canonical POS, or the *product of maxterms*, or the *maxterm expansion* form. In the SOP approach, the thought process involved in finding the terms that make the function $Z = 1$, so we basically defined true literals as 1 and complement literals as 0. In the POS approach, the thought process involves finding the terms that make the function $Z = 0$, so when writing the POS from the truth table, true literals are identified as 0 and complement literals are identified as 1. This change in the underlying thought process is why maxterm #0, or M0, is $(A + B + C)$, corresponding to the 000 row of the truth table. Take a minute to digest the ratiocination behind this convention.

Proof to show that $Z_{POS} = Z_{SOP}$:

$$Z(A, B, C) = \sum m(3, 5, 6, 7) = \overline{A}BC + A\overline{B}C + AB\overline{C} + ABC$$

$$\overline{\overline{Z}}(A, B, C) = \left(\overline{\overline{A}BC + A\overline{B}C + AB\overline{C} + ABC} \right)$$

$$= \overline{(A + \overline{B} + \overline{C}) (\overline{A} + B + \overline{C}) (\overline{A} + \overline{B} + C) (\overline{A} + \overline{B} + \overline{C})}$$

$$Z(A, B, C) = \overline{\overline{Z}}(A, B, C) = Z_{SOP} = \overline{M_3 M_5 M_6 M_7} = m_3 + m_5 + m_6 + m_7$$

$$Z_{POS} = M_0 M_1 M_2 M_4 = m_3 + m_5 + m_6 + m_7 = Z_{SOP}$$

After obtaining the canonical POS expression, step 4b, direct implementation of the canonical POS expression, will yield four three-input OR gates and one four-input AND gate. Notice the similarity to the canonical SOP implementation.

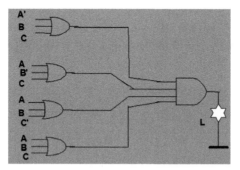

The canonical POS cost of implementation for this design is the same as the canonical SOP cost of implementation. You may verify this on your own.

After obtaining the gate implementation, proceed to step 5b, minimization of the canonical POS.

```
Z(A,B,C) = (A+B+C).(A+B+C').(A+B'+C).(A'+B+C)
Z(A,B,C) = [(A+B+C).(A+B+C')].[(A+B+C).(A+B'+C)].
                    [(A+B+C).(A'+B+C)]
[X = X .X,  Laws of Idempotence]

Z(A,B,C) = (A+B).(A+C).(B+C) [Laws of Complementarity,
(X+Y).(X+Y') = X]
```

For this design, the cost of implementation of the minimized POS is the same as the cost of the minimized SOP, as presented in Section 0. You may verify this yourself.

6.1.3 Self-duality

The majority gate function is a special function in Boolean algebra because it possesses the trait of self-duality. This statement essentially says that $Z_{MAJORITY}^{DUAL} = Z_{MAJORITY}$.

From the minimized POS expression, we obtain $Z^{\text{DUAL}}_{\text{MAJORITY}}$ to be

$$Z^{\text{DUAL}}(A, B, C) = A \cdot B + A \cdot C + B \cdot C$$

Remember the process to finding the dual of a Boolean function involves interchanging ANDs and ORs, 0s and 1s, and keeping the literals as given, that is, *don't remove NOTs*.

By expanding the minimized POS expression, we obtain Z_{MAJORITY} as

$$Z(A, B, C) = (AA + AC + BA + BC)(B + C)$$

$$= (A + AC + BA + BC)(B + C)$$

$$= [A(1 + C + B) + BC] \cdot (B + C) = (A + BC) \cdot (B + C)$$

$$Z(A, B, C) = AB + AC + BBC + BCC$$

$$= A \cdot B + A \cdot C + B \cdot C = Z^{\text{DUAL}}(A, B, C)$$

6.2 Why Different Representations: Two-Level Logic Implementation Styles

Why should we derive both SOP and POS representations? Is deriving one not good enough? Well, not exactly. Some logic representations are more efficient to implement than others, depending on the technology of interest. Depending on the company a circuit designer works for, he or she may design with POS expressions or SOP expressions or both.

For example, CMOS logic is usually focused on minimizing transistor count in order to reduce circuit area, so choosing to use AND and OR gates over NAND and NOR gates would not be an area-efficient design. Within the POS and SOP implementations for two-level logic, there exist eight different logic representations: four SOP representations and four POS representations. The representations are categorized as POS and SOP because the POS representations are derived from the POS expression and the SOP representations are derived from the SOP expression. The next two sections will derive the majority gate expression as each logic representation.

6.2.1 SOP representations

- **AND-OR**: $Z = Z_{\text{SOP}} = A \cdot B + B \cdot C + A \cdot C$

This implementation style is the SOP representation already derived.

- **NAND-NAND**: $Z = \overline{(\overline{Z})} = \overline{\overline{(A \cdot B + B \cdot C + A \cdot C)}}$
  $= \overline{(\overline{AB} \cdot \overline{BC} \cdot \overline{AC})}$

This implementation style is ideal for CMOS logic since NAND requires fewer transistors.

- **OR-NAND (inverted inputs)**: $Z = \overline{(\overline{AB} \cdot \overline{BC} \cdot \overline{AC})}$
  $= \overline{(\overline{A} + \overline{B}) \cdot (\overline{B} + \overline{C}) \cdot (\overline{A} + \overline{C})}$

This implementation style is good for printed circuit board (PCB) wiring and for wired logic, wired AND and wired OR.

- **NOR-OR (inverted inputs)**: $Z = \overline{(\overline{A} + \overline{B}) \cdot (\overline{B} + \overline{C}) \cdot (\overline{A} + \overline{C})} =$
  $\overline{(\overline{A} + \overline{B})} + \overline{(\overline{B} + \overline{C})} + \overline{(\overline{A} + \overline{C})}$

This implementation style is available for programmable logic arrays (PLAs), which are normally auto generated using computer-aided design (CAD) tools.

6.2.2 POS representations

- **OR-AND**: $Z = Z_{POS} = (A + B) \cdot (B + C) \cdot (A + C)$
- **NOR-NOR**: $Z = \overline{(\overline{Z})} = \overline{\left[\overline{(A + B) \cdot (B + C) \cdot (A + C)} \right]}$
  $= \overline{\left[\overline{(A + B)} + \overline{(B + C)} + \overline{(A + C)} \right]}$
- **AND-NOR (inverted inputs)**: $Z = \overline{\left[\overline{(A + B)} + \overline{(B + C)} + \overline{(A + C)} \right]}$
  $= \overline{\left[(\overline{A} \cdot \overline{B}) + (\overline{B} \cdot \overline{C}) + (\overline{A} \cdot \overline{C}) \right]}$
- **NAND-AND (inverted inputs)**: $Z = \overline{\left[(\overline{A} \cdot \overline{B}) + (\overline{B} \cdot \overline{C}) + (\overline{A} \cdot \overline{C}) \right]}$
  $= \overline{(\overline{A} \cdot \overline{B})} \cdot \overline{(\overline{B} \cdot \overline{C})} \cdot \overline{(\overline{A} \cdot \overline{C})}$

6.2.3 Compatible representations for CMOS design

This section deals with which logic representations are the easiest to convert between one another. By *easiest*, we mean the ability to swap out gates without altering the logic function itself. NAND-NAND and AND-OR are compatible, and OR-AND and NOR-NOR are also compatible, as shown in Figure 6.2 and Figure 6.3.

Note: All inputs must appear in the second level for this trick to work. If an input is introduced in the first level, then the trick does not work, as shown in Figure 6.4. You must invert the input, A in this case.

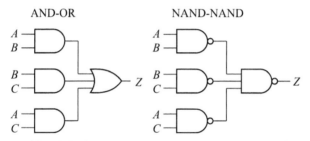

Figure 6.2 AND-OR to NAND-NAND conversion and vice versa. Interchange the gates and Z remains unchanged!

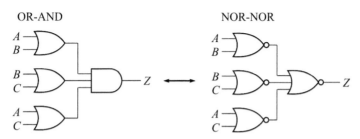

Figure 6.3 OR-AND to NOR-NOR conversion and vice versa. Interchange the gates and Z remains unchanged!

6.3 Hardware Description Languages

Hardware description languages (HDLs) are used to describe circuits in a similar way as computer programs. Sometimes being able to test/simulate a circuit's behavior before writing out truth tables and churning through Boolean algebra is very useful in designing against potential hazards. In industry, most designers do not tread through the full combinational logic

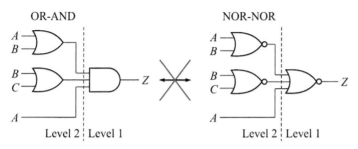

Figure 6.4 Incorrect conversion from OR-AND to NOR-NOR (all inputs must be introduced in Level 2).

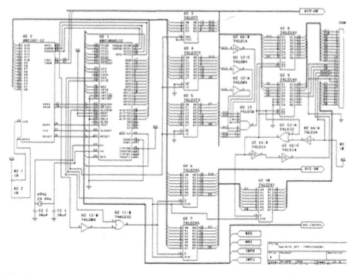

Figure 6.5   Complex, messy circuit schematic with extraneous information that reduces readability and understanding. Source: Timex/Sinclair Computers (timexsinclair.com).

design process delineated in this chapter. The circuit in Figure 6.5 provides a motivation as to why HDLs are important.

As shown in Figure 6.5, circuit schematics can contain extraneous information that hinders the understanding of a circuit that may perform a simple task. The level of detail in a schematic, showing wires and multiple bus lines, takes away from easily understanding the schematic diagram. In the 1980s, as circuit complexity grew from designing simple functions to interactions involving hundreds to thousands of gates, there came a need for a better way of describing large systems. Today, three popular HDLs are used in classrooms, the government, and industry: VHDL, Verilog, and SystemC.

A Verilog implementation will be used to bridge both the stairwell lamp problem and the majority gate design. Figure 6.6 shows a Verilog code that implements both designs. Both functions are contained in one module, and also entwined in this module is the testbench. The Verilog code will be described in detail in the next section.

6.3.1 Majority gate and stairwell lamp verilog implementation

The item **'timescale 1ns/1 ps** sets the simulator time reference and precision, that is, the simulator will reference all timed values in units of 1 ns and the smallest time step will be 1 ps. The line **module probs (majority, lamp_on)** defines the logic module named **probs** with input/output ports **majority** and **lamp_on**. The next line defines the direction of the variables **majority** and **lamp_on** as outputs stored in registers. The line after that defines three registers, A, B, and C.

Following the register declarations are three **always** blocks, which control the switching of the three registers, A, B, and C. C switches to its complement C' every 5 time units, B switches to its complement B' every 10 time units, and A switches to its complement A' every 20 time units. In 40 time units, A switches 2 times, B switches 4 times, and C switches 8 times. With this setup, we are able to generate all combinations of A, B, and C for a truth table representation. The time units here are referenced back to our timescale, which is 1 ns, so henceforth, nanoseconds will be used instead of time units.

After the register switching blocks, there's the expression **always @ (A or B or C)**. This expression signifies that any statements within this **always** block are re-evaluated whenever register A changes or B changes or C changes. Within this block are two condition statements that are the derived equations for the stairwell lamp problem and the majority gate example. If the evaluated expression is true, then the output is 1; if the evaluated expression is false, then the output is 0. These two conditional statements set the output variables **lamp_on** and **majority**.

The last block in the Verilog code of Figure 6.6 is the **initial** block, which is used as a start position during simulation, that is, timestep 0. The **$monitor** statement prints the string to the computer screen and is used specifically in this code to print out the register values for A, B, C, **majority**, and **lamp_on**. **A <= 0; B <= 0; C <= 0;** sets initial values (at timestep 0) for the A, B, and C registers, respectively. **#39; $finish;** instructs the simulation to end after a 39 ns delay.

Figure 6.7 shows the simulation results, and it looks exactly like the truth table representations for both the majority gate and stairwell lamp functions. Figure 6.8 presents the simulation results in graphical format.

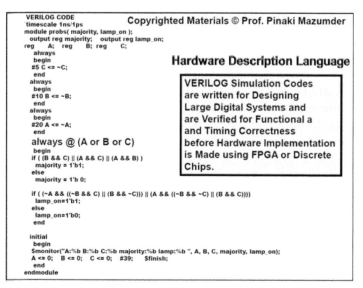

Figure 6.6   Verilog code implementing the lamp function and the majority gate function.

6.3.2 Full adder verilog implementation

The majority gate and stairwell lamp functions are actually functions used in a binary adder; Figure 6.9 describes the full adder (FA). The sum function, S, corresponds to the stairwell lamp results, while the carry-out function, C_o, corresponds to the majority gate function. The FA performs binary addition of two 1-bit numbers with a carry-in and produces a sum output and a carry-out.

Figure 6.10 shows a short Verilog module representation of the FA. There is no test bench in this Verilog code, only the FA description. Inputs are declared as A, B, and C_i, while outputs are declared as S and C_o. In addition to S and C_o being outputs, they are also stored on registers. In the **always** block, S and C_o are evaluated whenever any of the inputs changes.

6.3.3 Ripple carry adder

Figure 6.11 shows a *hierarchical* Verilog description of a 4-bit ripple carry adder (RCA). An RCA is implemented with multiple FAs, where the carry-out of one stage is the carry-in of the next stage. The RCA in this example is able to perform addition of two 4-bit numbers, X and Y. After implementing the FA module, multiple FA modules can be instantiated just like class

```
/usr/caen/ius-8.2/tools/bin/ncverilog    probs.v


ncsim> source /usr/caen/ius-8.2/tools/inca/files/ncsimrc
ncsim> run

A:0 B:0 C:0    majority:0      lamp:0
A:0 B:0 C:1    majority:0      lamp:1
A:0 B:1 C:0    majority:0      lamp:1
A:0 B:1 C:1    majority:1      lamp:0
A:1 B:0 C:0    majority:0      lamp:1
A:1 B:0 C:1    majority:1      lamp:0
A:1 B:1 C:0    majority:1      lamp:0
A:1 B:1 C:1    majority:1      lamp:1

Simulation complete via $finish(1) at time 39 NS + 0
./probs.v:46    $finish;

ncsim> exit
```

Figure 6.7   Verilog code simulation results showing the outputs for all input combinations.

instantiations in software programming. Four separate FAs are instantiated and the connections between them are achieved through wires named C_o0, C_o1, and C_o2. The wire connections are internal to the **RippleAdder4b** module and are not accessible outside this module.

6.4 Conclusion and Key Points

This chapter reinforced the design methodology involving SOP expressions and introduced an alternative methodology involving POS expressions through the design of a majority gate. With the SOP implementation, minterms were considered, but with the POS implementation, maxterms were introduced. The motivations behind both implementations were addressed, and eight logic representations derived from the canonical SOP and the canonical POS were presented. Lastly, a Verilog description of the majority gate function was presented, and two sample codes (FA and 4-bit RCA implementations) were used to introduce hierarchical Verilog descriptions.

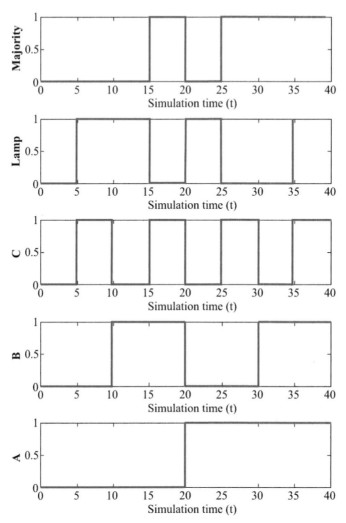

Figure 6.8   Verilog code simulation results shown graphically.

- Minterms depend on the logical rule that the combination of inputs will make the logic function's output true. Maxterms, on the other hand, depend on the combination of inputs making the logic function's output false. Minterms can be combined to describe an SOP implementation of the logic function, and maxterms can be combined to describe a POS implementation of the logic function.

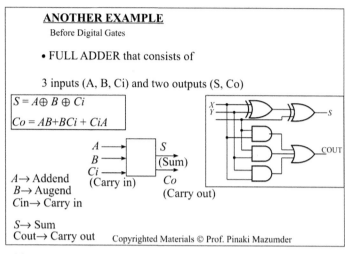

ANOTHER EXAMPLE

Before Digital Gates

• FULL ADDER that consists of

3 inputs (A, B, Ci) and two outputs (S, Co)

$S = A \oplus B \oplus Ci$

$Co = AB + BCi + CiA$

$A \rightarrow$ Addend
$B \rightarrow$ Augend
$Cin \rightarrow$ Carry in

$S \rightarrow$ Sum
$Cout \rightarrow$ Carry out

A — S (Sum)
B —
Ci — Co (Carry out)
(Carry in)

Copyrighted Materials © Prof. Pinaki Mazumder

Figure 6.9 Full adder implementation showing the sum and the carry-out functions.

Full adder (FA)

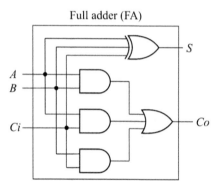

```
module FA (A, B, Ci, S, Co);
  input A, B, Ci;
  output reg S, Co;

  always @ (A or B or Ci)
  begin
    S <= A ^ B ^ Ci;
    Co <= (A & B) | (A & Ci) | (B & Ci);
  end
endmodule
```

Figure 6.10 Full adder module schematic and behavioral Verilog description.

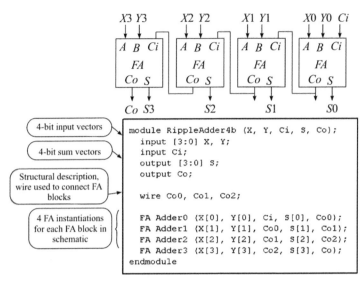

Figure 6.11   Hierarchical ripple carry adder structural Verilog implementation.

- There are functions that exhibit self-duality, and one example of such is the majority gate.
- Different representations (SOP, POS, etc.) of logic functions exist because some representations are easier to implement or utilize fewer hardware resources than others depending on technology used for implementation. For example, CMOS logic is usually focused on minimizing transistor count in order to reduce circuit area, so choosing to use AND and OR gates over NAND and NOR gates would not be an area-efficient design.
- NAND-NAND and AND-OR logic implementations are compatible, and OR-AND and NOR-NOR logic implementations are also compatible.
- HDLs describe circuits in a similar way as computer programs and allow for simulation of logic behavior for verification. Verilog implementations of majority gate design, full adder, and ripple carry adder are provided as examples of HDL implementations of logic functions. The ripple carry adder implementation shows that hierarchical description is supported in Verilog.

Problems on Combinational Circuits: Part II

1. A minority gate consists of three inputs A, B, and C and one output Z. The output $Z = 1$ if and only if two or more of the three inputs are 0.

   (a) Draw the truth table of the minority gate showing the relationship between the inputs A, B, and C and the output Z.

   (b) Write the canonical SOP and POS expressions for Z.

   (c) Minimize the SOP and POS expressions using Boolean algebra. Show all steps and state the Boolean laws required to minimize the expressions.

   (d) Is the minority gate logic function self-dual such that Z and its dual are identical?

   (e) Is the minority gate functionally complete in the sense that any arbitrary Boolean equation can be realized by using only minority gates? Prove your answer. Note that both NAND and NOR gates are individually functionally complete since you can construct primitive Boolean operations such as NOT, AND, and OR using only NAND/NOR gates.

2. A combinational circuit has four inputs W, X, Y, and Z and one output V. The output is 1 if and only if the 84-2-1binary-coded decimal (BCD)-coded digit represented by ABCD is less than 6.

   (a) Draw a truth table relating V with the four inputs. Note that input combinations corresponding to non-code-words never occur and you need not include them in your truth table.

   (b) Write V as a canonical sum of products, that is, a sum of minterms.

   (c) Minimize the canonical expression by using Boolean algebra and state all Boolean laws required in order to minimize the output equation.

3. A decade counter represents decimal digits 0 through 9 by the following a 4-bit (A, B, C, D) Gray code.

Decimal	a	b	c	d
0	0	0	0	0
1	0	0	0	1
2	0	0	1	1
3	0	0	1	0
4	0	1	1	0
5	0	1	1	1
6	0	1	0	1
7	0	1	0	0
8	1	1	0	0
9	1	0	0	0

We want to generate a function $z\ (a,b,c,d)$ for which $z = 1$ when the code represents an odd decimal digit, $z = 0$ when it represents an even decimal digit, and $z =$ "don't care" when it is an unspecified 4-bit value. Which of the following specifies the function? (Note: Δ denotes the list of "don't care.")

A. $z = \sum(1, 3, 5, 7, 9) + \Delta(10, 11, 12, 13, 14, 15)$
B. $z = \sum(1, 2, 4, 7, 8) + \Delta(9, 10, 11, 13, 14, 15)$
C. $z = \sum(0, 2, 4, 6, 8) + \Delta(10, 11, 12, 13, 14, 15)$
D. $z = \sum(0, 3, 5, 6, 9) + \Delta(10, 11, 12, 13, 14, 15)$
E. None of the above

4. A logic designer has come up with the following SOP expression for a function $z\ (a,b,c,d)$: $z(a, b, c, d) = c\bar{d} + abd + a\bar{b}\bar{c} + \bar{a}\bar{b}c\bar{d}$. The designer's boss says the expression is not minimal, and asks the designer to redo it and look for a less costly solution. Which of the following is a minimal expression that would fulfill the boss's request?

A. $z(a, b, c, d) = c\bar{d} + \bar{b}d + a\bar{b}\bar{c} + abc$
B. $z(a, b, c, d) = c\bar{d} + \bar{b}d + a\bar{c}d + abc$
C. $z(a, b, c, d) = c\bar{d} + \bar{b}d + abd + abc$
D. $z(a, b, c, d) = c\bar{d} + \bar{b}d + a\bar{b}\bar{c} + a\bar{c}d$
E. The original expression is minimal.

5. A Boolean network having two control inputs A and B connects the output Z to two data inputs V and W or connects the output to 0 or 1, depending on the values of the control inputs, A and B, as shown below. If $A = B = 1$, the output $Z = 1$. If $A = 1$ and $B = 0$, then $Z = V$. If $A = 0$ and $B = 1$, then $Z = W$.

Partial truth table of the problem:

Decimal	a	b	c	d
0	0	0	0	0
1	0	0	0	1
2	0	0	1	1
3	0	0	1	0
4	0	1	1	0
5	0	1	1	1
6	0	1	0	1
7	0	1	0	0
8	1	1	0	0
9	1	0	0	0

In the above table, the values of Z corresponding to P, Q, and R are:

A. $P = 1$, $Q = 0$, $R = 0$
B. $P = 0$, $Q = 1$, $R = 0$
C. $P = 0$, $Q = 0$, $R = 1$
D. $P = 0$, $Q = 0$, $R = 0$
E. None of the above

6. In a city council committee, there are 4 voting members to pass a resolution and a chairperson who votes only to break a tie to decide the outcome of a resolution under consideration. Assume that A, B, C, and D denote votes of the four voting members and E denotes the tie-breaking vote of the chairperson, while the outcome of the resolution is denoted by Z. If a Yes vote is indicated by 1 and a No vote by 0, write a truth table that reflects all possible outcomes of the resolution, where $Z = 1$ if the resolution passes. Note that the truth table should not include

all 32 entries as the value of E is Don't care, i.e. X, when the majority of voting members agree to pass or not pass the resolution. Design a Boolean circuit that performs the above function both in NAND-NAND form as well as in NOR-NOR form to compare which style of implementation reduces the number of transistors. Note that an M-input NOR gate contains 2M transistors.

7. Determine which of the Boolean expressions below are self-dual:

   (a) $f(a, b, c) = a'c' + a(b \odot c)$
   (b) $f(a, b, c) = \sum (0, 2, 4, 6)$
   (c) $f(a, b, c) = b'c' + a'(b \oplus c)$

8. Convert the following Boolean functions into NAND-NAND representations:

   (a) $f(a, b, c) = \sum (1, 3, 4, 6)$
   (b) $f(a, b, c) = (a \oplus c).b$
   (c) $f(a, b, c) = b'c' + a'b'$

9. Convert the following Boolean functions into OR-NAND and NOR-OR representations:

   (a) $f(a, b, c) = \sum (0, 1, 5, 7)$
   (b) $f(a, b, c) = b' + ac$
   (c) $f(a, b, c) = a.(b' \odot c)$

10. Convert the following Boolean functions into NOR-NOR representations:

   (a) $f(a, b, c) = \prod (1, 3, 4, 6)$
   (b) $f(a, b, c) = (a \oplus c).b$
   (c) $f(a, b, c, d) = (a'+c')(b+d')$

11. Convert the following Boolean functions into AND-NOR and NAND-AND representations:

   (a) $f(a, b, c, d) = \prod (1, 5, 6, 8, 11, 14)$
   (b) $f(a, b, c) = \sum (0, 3, 6, 7)$
   (c) $f(a, b, c) = a + b'c$

12. Generate Verilog codes for the tasks given below:

   (a) Design a Verilog module for a three-input majority gate.
   (b) Using the module you designed above, generate a hierarchical Verilog module that performs $f(a, b, c) = (a$ AND $b)$ OR $(b$ AND $c)$.

Combinational Logic Minimization

This chapter will introduce ways to minimize logic circuits using Karnaugh maps (K-maps). Three-variable and four-variable K-maps will be introduced. Several examples illustrating how to use K-maps will be presented.

Terms introduced in this chapter: Karnaugh map, implicant, prime implicant, essential prime implicant

Competency Objectives: At the end of this chapter, you will be able to:

1. Convert a truth table representation to a K-map representation, and minimize a Boolean expression using the K-map representation.
2. Determine minimized sum of product and product of sum expressions by selecting implicants containing minterms and maxterms, respectively.
3. Determine essential prime implicants and prime implicants and verify minimized Boolean expression obtained by K-maps.

7.1 Representation for Minimization: Summarization

From the previous chapter's majority gate, the truth table is represented here (Figure 7.1).

From the truth table, we can define input A as the most significant bit (MSB) and input C as the least significant bit (LSB). Each case presented in the truth table corresponds to a minterm and a maxterm. The truth table can

Input			Output (F)	
A	B	C		
0	0	0	0	m_0
0	0	1	0	m_1
0	1	0	0	m_2
0	1	1	1	m_3
1	0	0	0	m_4
1	0	1	1	m_5
1	1	0	1	m_6
1	1	1	1	m_7

Figure 7.1   Majority gate truth table with minterm identification.

be viewed as a table that either identifies cases for when the output F is 0 or identifies cases for when the output F is 1.

7.1.1 Intuitive design approach

This problem is simple enough for an intuitive design process, whereby two or more inputs equaling 1 need to cause the output to be 1, so we could have written ($A = 1$ AND $B = 1$) OR ($A = 1$ AND $C = 1$) OR ($B = 1$ AND $C = 1$). From this statement, the Boolean algebra expression can be obtained. If the intuitive design approach is not utilized, the Boolean expression can still be obtained from the truth table (Figure 7.1).

7.1.2 Boolean minimization

The truth table presents four cases where $F = 1$ and four cases where $F = 0$. Since the output F can be seen either as a summation of minterms or as a product of maxterms, we can write two Boolean expressions from the truth table and use Boolean algebra to simplify the expressions. These two expressions constitute two approaches that are appropriately named the *minterm approach* and the *maxterm approach*.

Minterm approach	Maxterm approach
$F = m_3 + m_5 + m_6 + m_7$ $= \bar{A}BC + A\bar{B}C + AB\bar{C} + ABC$ $= (\bar{A}BC + ABC) + (A\bar{B}C + ABC)$ $+(AB\bar{C} + ABC) \quad (\because X + X = X)$ $= BC + AC + AB \quad (\because X + \bar{X} = 1)$ NOTE : $m_3 = (A = 0)\&(B = 1)\&(C = 1)$ $m_3 = (\bar{A} = 1)\&(B = 1)\&(C = 1)$ $m_3 = \bar{A} \cdot B \cdot C = \bar{A}BC$	$F = M_0 \cdot M_1 \cdot M_2 \cdot M_4$ $= (A + B + C) \cdot \left(A + B + \overline{C}\right)$ $\quad \cdot \left(A + \overline{B} + C\right) \cdot \left(\overline{A} + B + C\right)$ $= \left[(A + B + C) \cdot (A + B + \overline{C})\right]$ $\quad \cdot \left[(A + B + C) \cdot (A + \overline{B} + C)\right]$ $\quad \cdot \left[(A + B + C) \cdot (\overline{A} + B + C)\right]$ $(\because X \cdot X = X)$ $= (A + B) \cdot (A + C) \cdot (B + C)$ $(\because X \cdot \overline{X} = 0)$

Through Boolean algebra, you can show that $(B \cdot C) + (A \cdot C) + (A \cdot B) = (A+B) \cdot (A + C) \cdot (B + C)$, so both minterm and maxterm approaches yield the same Boolean relationship. Although minterms and maxterms can be written and minimized using Boolean algebra, this approach is not the best way of minimizing. Due to Boolean algebra's nonsystematic way of minimizing, it may be a difficult or messy process to obtain a minimized expression when more variables are added to the mix. This difficulty of and need for a more systematic method leads to the graphical representation of the *Karnaugh map* (*K-map*).

7.2 Graphical Method: The Karnaugh Map

The K-map is a graphical method that facilitates systematic minimization of logic functions. Using the K-map, the minimized logic expression can be guaranteed, unlike a messy Boolean algebra derivation. The K-map is just another representation of the information in a truth table. Since truth tables can be viewed in terms of either outputs that equal 1 or outputs that equal 0, the K-map can also be viewed in the same way. Referring back to the majority gate example, the K-map in Figure 7.2 is constructed from the truth table in Figure 7.1.

In this K-map, the minterms from the truth table are transferred to the designated positions. From the truth table, input A is the MSB and input C is the LSB. The designated minterm positions follow this convention. This definition enables a systematic way to convert from truth table to a K-map. Notice that if the positions of any of the inputs are switched, for example, B becomes MSB and A becomes LSB, the output values of the corresponding minterms will differ in each position. Figure 7.2 also shows different regions around the K-map—the region where A is defined, B is defined, etc. This

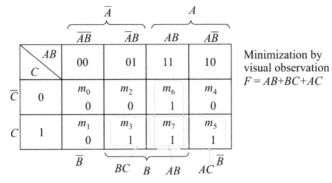

Minimization by
visual observation
$F = AB+BC+AC$

Figure 7.2 K-map representation from the SOP perspective. From the truth table translation, *A* is the MSB and *C* is the LSB. This arrangement produces the minterm pattern in the K-map. The 1's are grouped here, hence an SOP.

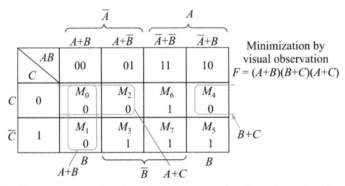

Minimization by
visual observation
$F = (A+B)(B+C)(A+C)$

Figure 7.3 K-map representation from the POS perspective. From the truth table translation, *A* is the MSB and *C* is the LSB. This arrangement produces the maxterm pattern in the K-map. The 0's are grouped here, hence a POS.

region-centric view results because the K-map is arranged in a way that similar literal definitions are adjacent to one another; that is why the top row reads 00, 01, 11, and then 10. These regions are from a sum of products (SOP) perspective, since the groupings shown in the K-map are groupings of 1's. When 1's are grouped together in powers of 2, an SOP expression can be written from the groupings. In Figure 7.2, three groupings (*AB*, *BC*, and *AC*) are made, and the region that contains an entire grouping denotes the product term. For example, the vertical grouping of m_6 and m_7 is contained entirely in *region A* and *region B* and partly in *region C* and *region $\overline{C}$*. Since *region A* and *region B* contain the entire grouping, the product term is just *AB*. Figure 7.3 shows an alternate view of the K-map.

In the K-map in Figure 7.2, we were interested in the input combinations that produced an output of 1, but sometimes we may be interested in outputs that equal to 0. When grouping 0's, the resultant expressions are POS expressions. From this perspective, we are interested in maxterms, hence the reversal of the regions in Figure 7.3. Take a minute to reconcile maxterm representations and minterm representations for one line in the truth table to see why this reversal is appropriate. Another feature of the K-map evident from Figure 7.3 is its toroidal structure: the K-map wraps around, as shown in the grouping of M_0 and M_4.

In the previous chapter, we showed that the majority function possesses the trait of self-duality. The dual expression of the SOP representation will yield the POS representation obtained in Figure 7.3. You may verify this.

The K-map is a graphical tool that does not provide any new information. It only enables a clearer view of the material already present in the truth table and what can be derived from Boolean algebra. In Figure 7.4, referring to M_3 and M_7, inputs B and C are both 1, while A changes from 0 to 1. From this observation, we can conclude that whenever B and C are 1, the output will be 1, independent of A. Similar conclusions can be made with M_5 and M_7 and with M_6 and M_7.

Still referring to Figure 7.4, applying Boolean algebra also yields similar conclusions with the combination of minterms m_3 and m_7, m_5 and m_7, and m_6 and m_7. The K-map essentially does not bring any new information to the table. The advantage the K-map possesses over other representations is that it presents the existing relationships in a way that is more apparent than scanning through the truth table. This advantage is amply observed in truth tables with larger groupings of 1's. The next section will provide examples of K-maps with three or four variables.

7.3 Three- and Four-Variable Karnaugh Maps for Logic Circuits

The three-variable K-map with minterm and maxterm positions has already been presented. Input A was the MSB and input C was the LSB, where as in the four-variable K-map, input A is the MSB and input D is the LSB. The four-variable K-map presented in Figure 7.5 is derived from a truth table with headers from the MSB to the LSB, *ABCD*. The arrangement on the four-variable K-map also allows the definition of the regions shown. The

A	B	C	Z	
0	0	0	0	M_0 Applying Boolean Algebra
0	0	1	0	M_1
0	0	0	0	M_2
0	1	1	1	M_3 → $(M3, M4) = BC$
1	1	0	0	M_4
1	0	1	1	M_5 → $(M5, M7) = CA$
1	1	0	1	M_6 → $(M6, M7) = AB$
1	1	1	1	M_7

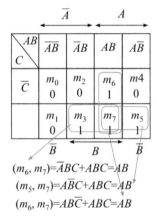

$(m_6, m_7) = \overline{A}BC + ABC = AB$

$(m_5, m_7) = A\overline{B}C + ABC = AB$

$(m_6, m_7) = AB\overline{C} + ABC = AB$

Figure 7.4   Connecting the minimization techniques of the K-map to Boolean algebra and the truth table.

demarcation for these regions is dependent on the arrangement 00, 01, 11, and 10 along both the top and the side of the K-map.

Analogous to truth table rows, each cell representation in the K-map represents either an AND gate or an OR gate. In the three-variable K-map, the cell representing the output due to m_0 can be thought of as $m_0 = \overline{ABC}$ or $M_0 = A + B + C$. In the three-variable case, each cell has three-input AND and OR gates, but in the four-variable case, each cell has four-input AND and OR gates. For example, in the four-variable case, m_5 corresponds to $\overline{A}B\overline{C}D$.

The larger the grouping or cluster on the K-map, the more literals that are eliminated from either the AND gate or the OR gate. From the K-map in

Figure 7.2, the grouping representing AC encompasses two minterms, m_7 and m_5. The grouping size (cluster size) that covers two minterms eliminates one literal—a size-2 cluster. For the three-variable case, instead of having a three-input AND gate, the size-2 cluster has a two-input AND gate. This literal elimination can be extended to larger clusters, as shown in the following examples (Figures 7.6, 7.7, and 7.8).

From the examples shown in Figure 7.6, the three-variable K-maps from left to right implement the functions $f = \sum(m_1, m_3, m_5, m_7) = C$ and $f = \sum(m_1, m_3, m_4, m_5, m_6, m_7) = A + C$, respectively. When the cluster size is 4 (a size-4 cluster), two literals are eliminated. So instead of a three-input AND gate, we have just noninverting gates that implement one literal. A size-4 cluster for a three-variable K-map could have also resulted with a grouping that provides a literal in complement form. This grouping would effectively also eliminate two literals, but instead of a noninverting gate, an inverting or NOT gate would be required. Figure 7.7 shows a size-4 cluster for a four-variable K-map.

The four-variable K-map starts out with each minterm composed of a four-input AND gate, so after two literals are eliminated, the result is a two-input AND gate. This is shown with BD, $\overline{BD}$, and CD selections presented in Figure 7.8. An extension of this to a cluster of size 8 (size-8 cluster) will yield an inverting or noninverting gate. Table 7.1 summarizes the relationship between cluster size and gate size for both three-variable and four-variable K-maps.

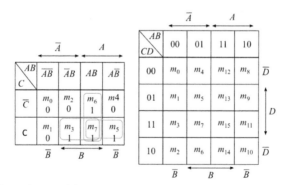

Figure 7.5   A three-variable and a four-variable K-map from an SOP perspective.

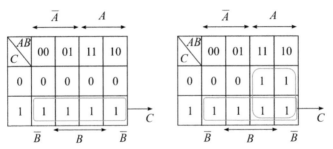

Figure 7.6 Three-variable K-map showing groupings of four 1's simplify the expression to one literal.

Table 7.1 Summary of cluster size and gate size relationship for three-variable and four-variable K-maps.

Three-variable K-map	Four-variable K-map
Cluster of size 2 → Two-input AND/OR gate	Cluster of size 2 → Three-input AND/OR gate
Cluster of size 4 → Inverting or noninverting gate	Cluster of size 4 → Two-input AND/OR gate
	Cluster of size 8 → Inverting or noninverting gate

7.4 Minimizing with Four-Variable K-maps

The K-map allows visual minimization of Boolean functions. Analogous to Boolean algebra, if minimization is not done systematically, there is room for error or for not arriving at a minimal representation of a function. A crucial advantage to a systematic method of minimization is the ability to

AB\CD	00	01	11	10
00	0	0	0	0
01	0	0	0	0
11	1	1	1	1
10	0	0	0	0

CD

AB\CD	00	01	11	10
00	1	1	1	1
01	0	0	0	0
11	0	0	0	0
10	1	1	1	1

BD

Figure 7.7 Four-variable K-map showing that groupings of four 1's simplify the expression to a two-input AND gate.

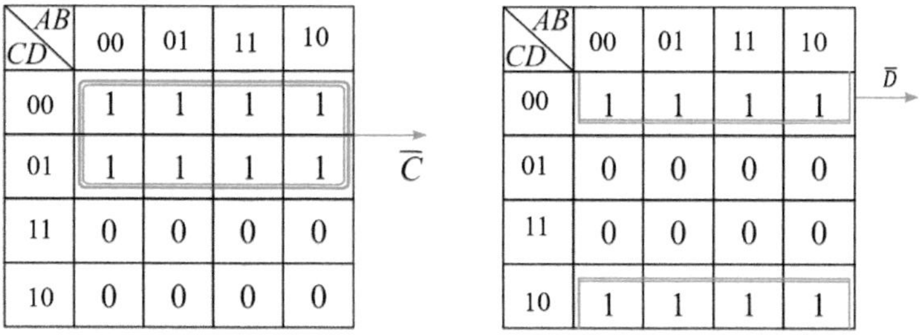

CD\AB	00	01	11	10
00	1	1	1	1
01	1	1	1	1
11	0	0	0	0
10	0	0	0	0

$\overline{C}$

CD\AB	00	01	11	10
00	1	1	1	1
01	0	0	0	0
11	0	0	0	0
10	1	1	1	1

$\overline{D}$

Figure 7.8   Four-variable K-map showing that groupings of eight 1's simplify the expression to one literal.

develop algorithmic and automated reduction. A systematic minimization with K-maps is a three-step process:

1. Identify the *essential prime implicants*.
2. Find the minterms (or maxterms) not included by the essential prime implicants.
3. Cover those minterms (or maxterms) using the minimum number of *prime implicants*.

7.4.1 Formal definitions

Any single 1 or a cluster of 1's (of size 2, 4, 8, etc.) that can be combined together on a map of the function F represents a product term that is called an *implicant* of F.

A product term implicant is called a *prime implicant* (*PI*) if it cannot be combined with another term to eliminate a variable.

A prime implicant is an *essential prime implicant* (*EPI*) to a function F if the prime implicant contains a minterm that is not covered by any other prime implicants of F.

Stepping away from the formal definitions, we will show through examples the meaning of these words and how to apply the three-step minimization process.

7.4.2 Example 1: detailed illustration of minimization

Problem: Find the minimum expression for the K-map in Figure 7.9. Distinguish between its EPIs and PIs.

CD \ AB	00	01	11	10
00	0	1	1	0
01	0	1	1	0
11	1	1	0	1
10	1	0	0	0

Figure 7.9   K-map problem for example 1.

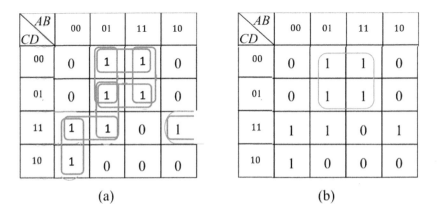

(a)   (b)

Figure 7.10   (a) K-map example showing all two-cluster implicants and (b) K-map example showing all four-cluster implicants.

Solution: From a minterm perspective, we can write an expression for the K-map: $F = \sum (2, 3, 4, 5, 7, 11, 12, 13)$.

Each of the minterms in F is an implicant. Wherever you have a 1 on the K-map, there exists a size-1 implicant.

Figure 7.9 contains eight implicants of size 1.

From the three-step process, step 1 involves finding the EPIs of the given problem. Size-1 implicants are not very interesting, and in this case they are not EPIs, as will be discovered when looking for size-2 implicants. Figure 7.10a shows the groupings when considering size-2 implicants.

There are eight implicants of size 2: $\overline{A}B\overline{C}$, $AB\overline{C}$, $B\overline{C}D$, $B\overline{C}\overline{D}$, $\overline{A}BD$, $\overline{A}CD$, $\overline{A}BC$, and $\overline{B}CD$. Can we go higher than size-2 implicants? Yes! There is one implicant of size 4 ($B\overline{C}$), as shown in Figure 7.10b. This is the

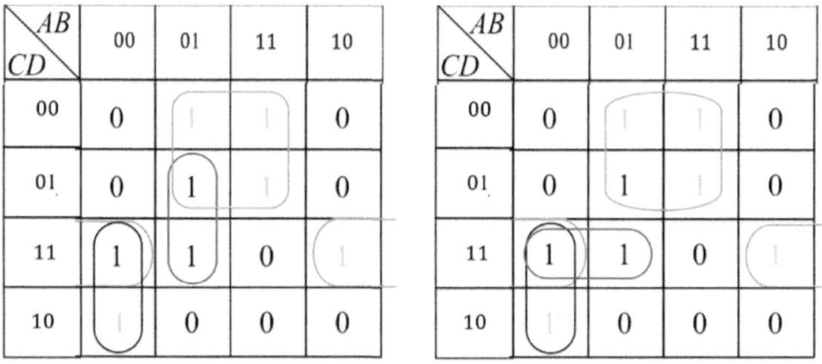

(a)   (b)

Figure 7.11 Example of K-map showing two different minimum representations. The green colored 1 in a PI is a distinguished minterm, making the PI into essential PI (EPI).

largest-size implicant we can obtain for F. This size-4 implicant contains four size-2 implicants: $\overline{A}B\overline{C}$, $AB\overline{C}$, $B\overline{C}D$, and $B\overline{C}\overline{D}$. These four implicants are automatically disqualified as PIs since they can be extended to form a larger implicant.

We are now left with four size-2 implicants and one size-4 implicant. The implicants left cannot be combined into a larger implicant; therefore these implicants are PIs. The PIs for this example are therefore $\overline{A}BD$, $\overline{A}CD$, $\overline{A}BC$, $\overline{B}CD$, and $B\overline{C}$. These PIs can be configured as shown in Figure 7.11 to cover all 1's in the K-map.

Figure 7.11 shows two K-maps with five large green 1's. These 1's represent minterms that can only be covered by one PI. Since only one PI can cover these minterms, those PIs are called EPIs. The EPIs for the current example are $\overline{A}BC$, $\overline{B}CD$, and $B\overline{C}$. Step 1 concluded!

Step 2 involves finding the terms not covered by an EPI, and in this example, only $\overline{A}BCD$ is not covered by an EPI. Step 3 involves using PIs to cover terms not covered by EPIs, and this is shown in two ways depicted in Figure 7.10a,b. Hence, the minimum expression can be one of two, that is,

$$F = B\overline{C} + \overline{A}BC + \overline{B}CD + \overline{A}\,B\,D$$

OR

$$F = B\overline{C} + \overline{A}BC + \overline{B}CD + \overline{A}\,C\,D$$

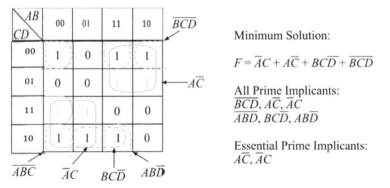

Minimum Solution:

$$F = \overline{A}C + A\overline{C} + BC\overline{D} + \overline{BCD}$$

All Prime Implicants:
$\overline{BCD}, A\overline{C}, \overline{A}C$
$\overline{A}B\overline{D}, BC\overline{D}, AB\overline{D}$

Essential Prime Implicants:
$A\overline{C}, \overline{A}C$

Figure 7.12   Example 2's K-map showing different implicants and minimum solution.

7.4.3 Example 2: prime implicant definition reinforcement

Problem: Find the minimum expression for the K-map in Figure 7.12. List its EPIs and PIs.

Solution: To find the EPIs, find the PIs in the problem by trying to form the largest groupings that are not fully contained within another grouping. This process results in two size-4 implicants ($\overline{A}C$ and $A\overline{C}$) and four size-2 implicants ($\overline{A}B\overline{D}$, $\overline{BCD}$, $AB\overline{D}$ and $BC\overline{D}$). These six implicants are the PIs.

Note that this problem is interesting because all the size-2 PIs share all their minterms with other PIs. These size-2 implicants are PIs because no other implicant supersedes these implicants, that is, these implicants are not a subset of any other implicants.

From Figure 7.12, only the two minterms labeled in large green 1's are covered by a single PI. Hence the PIs that cover these terms are EPIs. The EPIs for this example are $\overline{A}C$ and $A\overline{C}$. After covering the minterms with the EPIs, two minterms, $\overline{ABCD}$ and $ABC\overline{D}$, remain uncovered. These minterms can be covered with at least two PIs such as $\overline{A}B\overline{D}$ and $BC\overline{D}$. Step 3 of the three-step minimization process requires using the minimum number of PIs to cover the uncovered minterms.

7.4.4 Example 3: dealing with "Don't cares"

Figure 7.13 shows two examples of K-maps with "don't care" outputs denoted as X. "Don't cares" arise when the output of a certain input combination is not important to the desired functionality. One K-map deals

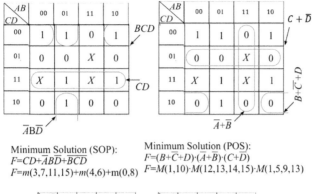

Minimum Solution (SOP):
$F=CD+\overline{A}B\overline{D}+\overline{B}C\overline{D}$
$F=m(3,7,11,15)+m(4,6)+m(0,8)$

Minimum Solution (POS):
$F=(B+\overline{C}+D)\cdot(\overline{A}+B)\cdot(C+D)$
$F=M(1,10)\cdot M(12,13,14,15)\cdot M(1,5,9,13)$

Figure 7.13 Example 3's K-map showing different implicants and multiple ways of approaching the problem.

with "don't cares" from an SOP perspective by grouping 1's, while the other deals with "don't cares" from a POS perspective by grouping 0's. "Don't cares" can either be treated as 1's or 0's, depending on what grouping produces the largest implicant. For the moment, ignore the two smaller K-maps illustrated in the example.

In Figure 7.13, each K-map representation can produce either a sum of minterms or a product of maxterms. The minterms that correspond to "don't cares" are identified by *d* instead of *m*, as given in the example. From this perspective, *D* is viewed as the LSB and *A* is the MSB.

In the K-map to the left, the inclusion of the *X* allowed a larger clustering than otherwise. When dealing with *X*'s, only cover the ones that allow better minimization. The covered *X* in this K-map is treated as 1 and the uncovered *X*'s are treated as 0's. This perspective is shown in the smaller K-map on the bottom based on the implicants used to cover the 1's and *X*'s. When determining EPIs, look for 1's that can only be covered by one PI. *Note: X's that are covered by only one PI do not make that PI an EPI; only 1's can make a PI an EPI.*

Given
$F = \sum m(3,4,6,7,15) + \sum d(2,9,11)$

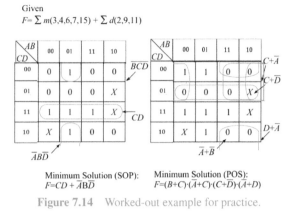

Minimum Solution (SOP):   Minimum Solution (POS):
$F = CD + \overline{A}B\overline{D}$ $F = (B+C)\cdot(\overline{A}+C)\cdot(C+\overline{D})\cdot(\overline{A}+D)$

Figure 7.14   Worked-out example for practice.

The EPIs in the K-map to the left are $\overline{BCD}$ and CD because of m_8 and m_{11}, respectively. For step 2, m_4 and m_6 are left uncovered by EPIs so $\overline{A}B\overline{D}$ is used to cover these remaining 1's. By using this current grouping, we are essentially casting our original function F into the smaller K-map, changing two X's to 1's and one X to 0.

The same problem can be viewed in a product of sums (POS) perspective by grouping 0's. The same procedure can be applied using steps 1 to 3. Still referring to Figure 7.13, the K-map to the right has the following PIs: $(C + \overline{D})$, $(\overline{A}+\overline{B})$, $(A+B+\overline{C})$, $(A+B+\overline{D})$, $(\overline{A}+\overline{C}+D)$, and $(B+\overline{C}+D)$. The EPIs are determined from the maxterms $(A+\overline{B}+C+\overline{D})$, $(\overline{A}+B+C+\overline{D})$, and $(\overline{A}+\overline{B}+C+D)$ since these terms can only be covered by one PI. Therefore the EPIs are $(C + \overline{D})$ and $(\overline{A} + \overline{B})$. After covering these terms with the EPIs, $(\overline{A}+B+\overline{C}+D)$ and $(A+B+\overline{C}+D)$ are left uncovered. Since the PI $(B + \overline{C} + D)$ can be used to cover the uncovered maxterms, it is chosen in order to obtain the minimum expression.

The small K-maps in Figure 7.13 shows that the two approaches do not necessarily produce the same results. The underlined output values show that in one K-map, the X is treated as a 1 and in the other it is treated as a 0. Depending on the logic implementation fabric, one approach may be chosen over the other, as discussed in logic representations in Chapter 6. Additionally, Figure 7.14 provides another example for extra practice. Can you find which X's are viewed differently between the two approaches?

7.5 Conclusion and Key Points

K-maps are used to minimize logic functions. Analogous to the truth table, the K-map can be viewed from a POS perspective or from an SOP perspective. Examples are used to explain the meanings of implicants, PIs, and EPIs.

- Boolean functions can be represented using a truth table, and truth tables can be drawn as K-Maps. Combining minterms or maxterms in K-Maps to form larger minterm or maxterm groupings eliminates literals.
- Systematic minimization of a Boolean function using K-Maps involves finding essential prime implicants, then finding minterm or maxterms that are not covered by the essential prime implicants and covering them with a minimum number of prime implicants.
- Don't cares can take a value of either a 0 or a 1. Don't cares, when included as a minterm grouping for minimizing a Boolean expression take on a value of 1. Don't cares, when included as a maxterm grouping for minimizing a Boolean expression take on a value of 0.

Appendix of K-Map

The Karnaugh map is a graphical technique that is used by logic designers to manually reduce the Boolean expression of a function. Usually at most 6-variable functions can be minimized using the K-map. The key in using K-maps is to combine 1's (alternatively 0's) to form the largest clusters of power of 2 that are defined as primary implicants (PIs). The next step is to identify all distinguished minterms in a PI that are not shared by other PIs; such a PI with one or more distinguished minterms is defined as an essential prime implicant (EPI). Therefore, in order to minimize a function, at first EPI's must be identified and all the minterms included in those EPI's will be automatically covered in the minimized Boolean expression for the function that must contain all the EPI's. Then, in order to write the minimized Boolean expression, one needs to include minimum number of additional PI's to cover the rest of the minterms in the original expression of the function. From the examples in this chapter, it is quite clear that the procedure can be mentally applied to write the minimized expression of a 4-variable function. In the following diagrams, 2 through 6 variable K-maps are drawn to identify the boxes for different minterms.

Two variable (AB) K-map

B \ A	0	1
0	m_0	m_2
1	m_1	m_3

Three variable (ABC) K-map

BC \ A	0	1
00	m_0	m_4
01	m_1	m_5
11	m_3	m_7
10	m_2	m_6

(A is MSB, and C is LSB)

Four variable (ABCD) K-map

CD \ AB	00	01	11	10
00	m_0	m_4	m_{12}	m_8
01	m_1	m_5	m_{13}	m_9
11	m_3	m_7	m_{15}	m_{11}
10	m_2	m_6	m_{14}	m_{10}

(A is MSB, and D is LSB)

Six Variable (ABCDEF) K-map

$\bar{A}$

EF \ CD	00	01	11	10
00	m_0	m_4	m_{12}	m_8
01	m_1	m_5	m_{13}	m_9
11	m_3	m_7	m_{15}	m_{11}
10	m_2	m_6	m_{14}	m_{10}

B (left block)

A

EF \ CD	00	01	11	10
00	m_{32}	m_{36}	m_{44}	m_{40}
01	m_{33}	m_{37}	m_{45}	m_{41}
11	m_{35}	m_{39}	m_{47}	m_{43}
10	m_{34}	m_{38}	m_{46}	m_{42}

$\bar{B}$ (lower blocks)

EF \ CD	00	01	11	10
00	m_{16}	m_{20}	m_{28}	m_{24}
01	m_{17}	m_{21}	m_{29}	m_{25}
11	m_{19}	m_{23}	m_{31}	m_{27}
10	m_{18}	m_{22}	m_{30}	m_{26}

EF \ CD	00	01	11	10
00	m_{48}	m_{52}	m_{60}	m_{56}
01	m_{49}	m_{53}	m_{61}	m_{57}
11	m_{51}	m_{55}	m_{63}	m_{59}
10	m_{50}	m_{54}	m_{62}	m_{58}

(A is MSB, and F is LSB)

Five variable (ABCDE) K-map

$\bar{A}$

DE \ BC	00	01	11	10
00	m_0	m_4	m_{12}	m_8
01	m_1	m_5	m_{13}	m_9
11	m_3	m_7	m_{15}	m_{11}
10	m_2	m_6	m_{14}	m_{10}

A

DE \ BC	00	01	11	10
00	m_{16}	m_{20}	m_{28}	m_{24}
01	m_{17}	m_{21}	m_{29}	m_{25}
11	m_{19}	m_{23}	m_{31}	m_{27}
10	m_{18}	m_{22}	m_{30}	m_{26}

(A is MSB, and E is LSB)

Note that if in a K-map $m_p = m_q$, then m_p and m_q can be combined together to form an implicant of size 2 if and only if $|p - q| = 1, 2, 4, 8, \ldots, 2^k$, because in that case those two minterms are adjacent on the K-map. One can, systematically, combine minterms to form larger clusters and identify the prime implicants (PI's) like in the 4-variable K-map examples discussed before.

However, the Karnaugh map (K-map) method becomes quite cumbersome as the number of variables increases beyond five. For Boolean expressions involving more than five Boolean variables, a systematic tabular method, called the *Quine–McCluskey* (Q-M) *method* will be discussed. The Q-M method was originally introduced by Willard Van Orman Quine and Edward McCluskey in the 1950s for determining the minimum sum-of-products (SOP) expression of a Boolean function. The systematic implementation of the Q-M method also allows for the use of writing simple computer programs that can automatically perform the minimization of Boolean expressions in a large digital network. The Q-M method will be covered in Chapter 20 after we introduce more important design methodologies of combinational and sequential digital networks where the K-map method is paramount. It must be emphasized here that the K-map is quintessential in logic design of subsequent chapters and it is covered up front, while the Q-M method is deferred to allow students to learn design techniques that they will apply in their laboratory experiments.

Problems on Karnaugh Map Minimization

1. Simplify the following Boolean functions using the K-map method:
   (a) $f(a, b, c) = \sum(0, 1, 3, 4)$
   (b) $f(a, b, c) = \sum(0, 2, 5, 7)$
   (c) $f(a, b, c) = \sum(1, 3, 4, 6)$

2. Simplify the following Boolean functions using the K-map method:
   (a) $f(a, b, c, d) = \sum(1, 3, 7, 11, 14, 15)$
   (b) $f(a, b, c, d) = \sum(0, 2, 6, 8, 9, 10, 13)$
   (c) $f(a, b, c, d) = \sum(1, 3, 7, 11, 13, 15)$

3. Simplify the following Boolean functions using the K-map method:
   (a) $f(a, b, c, d) = \sum(0, 1, 3, 6) + d(14, 15)$
   (b) $f(a, b, c, d) = \sum(4, 6, 9, 11) + d(3, 13)$
   (c) $f(a, b, c, d) = \sum(1, 5, 7, 13, 15) + d(3, 9, 14)$

4. Simplify the following Boolean functions using the K-map method:
   (a) $f(a, b, c, d) = \prod(0, 1, 3, 6) . D(14, 15)$
   (b) $f(a, b, c, d) = \prod(4, 6, 9, 11) . D(3, 13)$
   (c) $f(a, b, c, d) = \prod(1, 5, 7, 13, 15) . D(3, 9, 14)$

5. For the following K-map, a minimal SOP solution for the function is:

ab \ cd	00	01	11	10
00	1	1	1	1
01	0	0	0	d
11	0	0	1	d
10	1	1	1	1

   A. $\bar{a}c + c\bar{d} + \bar{b}c + \bar{b}\bar{c}$
   B. $\bar{b} + c\bar{d} + ac$
   C. $(\bar{b} + c)(\bar{a} + \bar{b} + \bar{d})$
   D. $b\bar{c} + abd$
   E. None of the above

6. The essential PI<u>s</u> of the function $f(a,b,c,d)$ whose K-map (blank cells are 1) is given below are:

ab

	00	01	11	10
00			0	0
01				0
11	0	0		0
10		0		

cd

A. $a\bar{c}\bar{d}, \bar{a}bc$
B. $a + \bar{c} + \bar{d}, \bar{a} + b + c$
C. $\bar{a} + c + d, a + \bar{b} + \bar{c}$
D. $\bar{a} + b + c, \bar{a} + b + \bar{d}, \bar{a} + c + d, b + \bar{c} + \bar{d}, a + \bar{c} + \bar{d}, a + \bar{b} + \bar{c}$
E. None of the above

7. Minimize the Boolean function represented in the following K-map by grouping the maxterms together (blank cells are 0).

ab \ cd	00	01	11	10
00	1			1
01	1		1	
11		1	1	
10	1			1

8. Find the PIs, EPIs, and the minimized expression represented in the following K-map.

ab \ cd	00	01	11	10
00	0	1	d	1
01	1	0	1	0
11	0	d	1	0
10	1	0	1	1

9. Find the PIs, EPIs, and the minimized expression represented in the following K-map.

ab \ cd	00	01	11	10
00	1	0	1	d
01	0	1	0	1
11	d	d	1	0
10	d	0	0	1

10. What are the relationships between F_1, F_2, and F_3 in the following K-maps?
    (Hint: Find the expression for F_1, F_2, and F_3.)

F_1

CD \ B	00	01	11	10
00	1	0	0	1
01	0	1	1	0
11	0	1	1	0
10	1	0	0	1

F_2

D \ AB	00	01	11	10
0	1	0	0	1
1	0	1	1	0

F_3

D \ B	0	1
0	1	0
1	0	1

A. $F_2 = F_3$ and $F_1 = \overline{F_2}$
B. $F_1 = F_3$ and $F_2 = \overline{F_3}$
C. $F_1 = F_3$ and $F_2 = \overline{F_1}$
D. $F_1 = F_2 = F_3 = B \odot D$
E. None of the above

11. Given $F(A, B, C, D) = \sum m(0, 1, 4, 6, 10) + \sum d(2, 5, 12, 15)$, using three two-input NOR gates and one three-input NOR gate, the above function can be realized. Assume the inputs of three two-input NOR gates are denoted by a, b, c, d, e, and f, respectively. Find which of the following input assignments of the NOR-NOR implementation of $F(A,B,C,D)$ is true.
    A. $a = A, b = B, c = \overline{A}, d = \overline{B}, e = \overline{C}, f = \overline{D}$
    B. $a = \overline{A}, b = \overline{B}, c = A, d = C, e = \overline{D}, f = \overline{C}$

C. $a = \overline{A}, b = \overline{B}, c = \overline{A}, d = C, e = \overline{D}, f = \overline{C}$
D. $a = \overline{D}, b = \overline{C}, c = \overline{A}, d = C, e = A, f = C$
E. None of the above

12. Generate a 2 × 3 (five-variable) K-map to minimize the following expressions:

a) $f(a, b, c, d, e) = \sum(1, 3, 7, \ 11, 17, 21, 25, 29, 31) + d(6, 9, 18, 26)$
b) $f(a, b, c, d, e) = \sum(0, 2, 4, 8, 13, 15, 19, 23, 28, 30) + d(6, 14, 18, 29)$
c) $f(a, b, c, d, e) = \sum(3, 5, 6, 9, 12, 16, 21, 26, 27) + d(0, 11, 19, 29, 31)$

13. Write the truth table of a binary-to-excess-3 BCD code converter. Using the K-map minimization technique, write the output variables as functions of input variables in minimized SOP form. Note that the output variables are "don't cares" (X's) for all invalid input patterns. Show the SOP in NAND-NAND form and then implement the above converter using **only** the minimum number of two-input NAND gates. How many 74LS00 chips are required to implement the converter?

8

Combinational Building Blocks

This chapter will introduce decoders and multiplexers and show how these components can be implemented using logic gates. The chapter also shows how Boolean functions can be implemented using multiplexers and decoders. Afterward some MSI integrated circuits (ICs) are introduced, specifically decoders, three-state buffers, encoders, multiplexers, parity circuits, and, lastly, comparison circuits. The section on MSI building blocks is more applicable to the lab where students work with gate-level ICs to implement certain functionalities.

Terms introduced in this chapter: binary decoder, multiplexers, binary encoder, priority encoder, even parity, odd parity

Competency Objectives: At the end of this chapter, you will be able to:

1. Build decoders and multiplexers in a hierarchical manner or in a single level using logic gates.
2. Use decoders, and multiplexers to implement Boolean functions.
3. Implement muxes, decoders, tri-state buffers, and encoders in Verilog.
4. Build a magnitude comparator for checking equality, and inequality relationships.

8.1 Decoders

The decoder is a logic circuit that converts coded inputs into coded outputs, whereby the input and output codes are different from one another. The

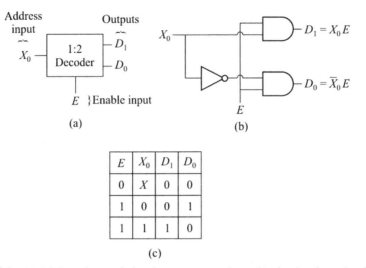

Figure 8.1   (a) 1:2 Decoder symbol and output expressions, (b) circuit schematic of a 1:2 decoder, and (c) truth table.

decoder encountered most often is the n-to-2^n decoder, or *binary decoder*. These decoders are covered in detail in this section. Figure 8.1a depicts a 1:2 decoder, which is used in this section to describe the functionality of a binary decoder. Figure 8.1b is the realization of the 1:2 decoder with logic gates, and Figure 8.1c is the truth table associated with the 1:2 decoder. X_0 is the input to the decoder (represents a 1-bit input). D_0 and D_1 are the outputs of the decoder. E is called the enable input and is essentially a signal that activates the decoder for proper operation.

The 1:2 decoder is the simplest decoder. From the gate-level depiction given, it is self-evident that when the decoder is enabled, only one output is active at a time. The 1:2 decoder has two outputs, so either D_1 is active or D_0 is active, but both D_1 and D_0 cannot be 1 at the same time. Each output corresponds to a minterm (or maxterm), depending on the input combination. The general form of a decoder is n-to-2^n, such as a 2:4 decoder, a 3:8 decoder, a 4:16 decoder, a 5:32 decoder, and so on.

8.1.1 Implementation of larger-bit decoders

Decoders can be implemented using logic gates or can be combined to create larger decoders. For example, the 2:4 decoder presented in Figure 8.2a could be implemented with AND gates by implementing each minterm with

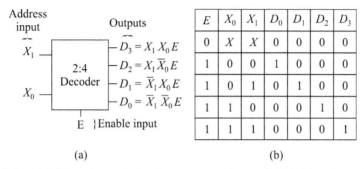

Figure 8.2 (a) 2:4 Decoder symbol and output expressions and (b) 2:4 decoder truth table.

an AND gate. This gate implementation of the 2:4 decoder is shown in Figure 8.3b. Again, the enable signal E is included because the decoder should not produce any ACTIVE outputs when not enabled. The enable signal is essentially a control signal that when its value is 0, all outputs are 0, as shown in the truth table in Figure 8.2b. *Note: The truth table uses X as a "don't care" and X_0 and X_1 as inputs.*

The hierarchical method of getting a 2:4 decoder is shown in Figure 8.3a. This example uses three 1:2 decoders and produces the same output as the AND gate implementation. Tracing through the hierarchical implementation, the first level is used as enable signals for the second level. When the first level is enabled, the Boolean value of X_1 is used to determine which decoder in the second level is to be enabled. From the truth table in Figure 8.2b, we see that if X_1 is 0, then $D_0(D_1)$ will be active if $X_0 = 0$ ($X_0 = 1$). Conversely, if X_1 is 1, then $D_2(D_3)$ will be active if $X_0 = 0$ ($X_0 = 1$). Hence, X_0 is used as input in the second level decoders in Figure 8.3a to decide how to name the outputs corresponding to the appropriate behavior of the expected 2:4 decoder. So the naming of the outputs follows that D_0 is at the top and D_3 is at the bottom.

Over-the-counter commercial decoder integrated circuits (ICs) are available, and some examples are the 74154 → 4:16 Decoder, the 74155 → 2:4 Decoder, and the 74141 → 4:10 Decoder, BCD to decimal.

Similar to the example shown in Figure 8.3a, smaller-bit decoders can be combined in a hierarchical manner to make larger-bit decoders. Therefore, the over-the-counter medium-scale integration (MSI) chips listed here can be combined to make larger-bit decoders, if necessary.

8.1.2 Using decoders to implement boolean functions

As seen in the introduction to decoders, decoders implement minterms. Hence they can be used to implement product of sums (POS) representations or sum or products (SOP) representations of Boolean functions with the aid of other logic gates. This section will exemplify how this can be accomplished.

8.1.2.1 Example 1

Problem: Implement the function $L(A,B,C) = \Sigma(1,2,4,7)$ using a 3:8 decoder, an inverter, and OR gates.

Solution: In this instance, A is the most significant bit (MSB) and C is the least significant bit (LSB). Since the decoder can be used to provide the minterms directly, all that is necessary is to combine the minterms with an OR gate. The enable signal is connected to a Logic 1 source—in this case a 5 V power supply.

8.1.2.2 Example 2

Problem: Implement the function $L(A,B,C,D) = \Pi(1,5,6,11,12,15)$ using a 3:8 decoder, an inverter, and OR gates.

Solution 1: This problem will be solved in two different ways: The first method is labeled "Solution 1," while the second method is "Solution 2." In this instance, A is the MSB and D is the LSB. From the Example 1 solution, we see that the decoder can be used to provide the minterms directly. In this case, we are concerned though with the product of maxterms. The next step is to convert the product of maxterms into sums of minterms in order to be

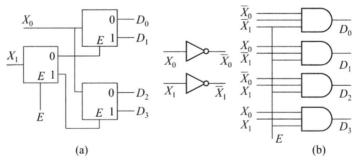

(a)   (b)

Figure 8.3 (a) Hierarchical design of a 2:4 decoder using 1:2 decoders and (b) 2:4 decoder implementation with AND gates.

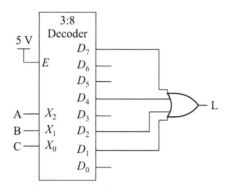

Figure 8.4 Circuit implementation of the function L(A,B,C) given in Example 1.

appropriate for decoder implementation.

$$L(A, B, C, D) = \Pi(1, 5, 6, 11, 12, 15) = \Sigma(0, 2, 3, 4, 7, 8, 9, 10, 13, 14)$$

Since the problem does not allow the use of a 4:16 decoder, we cannot implement this function with just one 3:8 decoder. We need two 3:8 decoders to realize this function. The minterms that we need to implement are therefore m_0, m_2, m_3, m_4, m_7, m_8, m_9, m_{10}, m_{13}, and m_{14}. We notice that minterms m_0 through m_7 can be implemented when $A = 0$ and minterms m_8 through m_{15} can be implemented with $A = 1$. We will implement m_0 through m_7 with one decoder and m_8 through m_{15} with another decoder. In addition, we will use $A = 1$ to select one decoder and $A = 0$ to select the other decoder; hence the enable signals for both inverters will be derived from the MSB A. The solution is presented in Scheme 8.1. The minterms implemented are identified.

Solution 2: Reiterating, A is the MSB and D is the LSB. Using DeMorgan's laws we can convert the product of maxterms into the sum of minterms.

$$L(A, B, C, D) = \prod(1, 5, 6, 11, 12, 15) = \left(\overline{\prod(1, 5, 6, 11, 12, 15)}\right)$$

$$= \left(\overline{\sum(1, 5, 6, 11, 12, 15)}\right)$$

The product of the maxterms M_1, M_5, M_6, M_{11}, M_{12}, and M_{15} is essentially the inverted output of the sum of the minterms m_1, m_5, m_6, m_{11}, m_{12}, and m_{15}. The solution is, therefore, presented in Scheme 8.2.

The process of using input signals through a decoder to obtain an output is the underlying idea behind read-only memory (ROM). Decoders are very important in the implementation of memory, especially ROM.

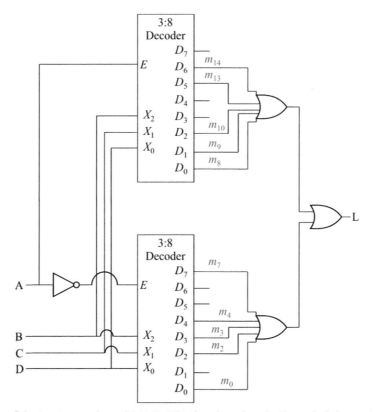

Scheme 8.1   Implementation of L(A,B,C,D) function given in Example-2 by tracing the minterms (solution-1).

The decoder takes a few inputs and creates a large number of outputs. We will next look at a logic device that takes multiple inputs and creates one output.

8.2 Multiplexers

Multiplexers, or MUXes, were briefly mentioned earlier when timing diagrams were introduced. They were previously used to explain the problems associated with timing hazards. A MUX is a multiple-input, single-output logic circuit that selects one path out of many on the basis of control signals. It differs from the binary decoder, because the binary decoder takes an *n*-bit input and creates a 2^n-bit output. The MUX, on the other hand,

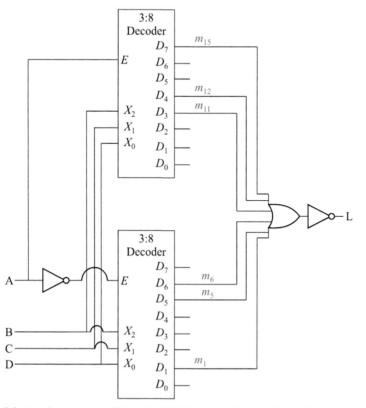

Scheme 8.2 Implementation of L (A,B,C,D) function given in Example-2 by tracing the maxterms (solution-2).

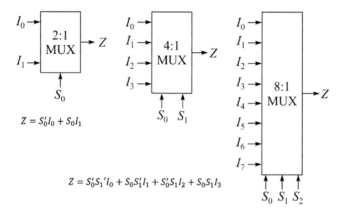

Figure 8.4 Multiplexer examples. From left to right: 2:1 MUX, 4:1 MUX, and 8:1 MUX.

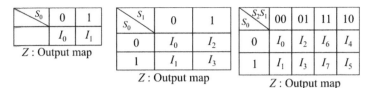

Figure 8.5   Multiplexer K-maps. From left to right: 2:1 MUX, 4:1 MUX, and 8:1 MUX.

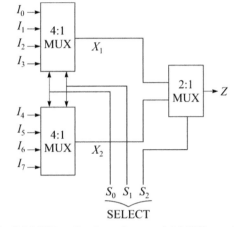

Figure 8.6   8:1 MUX realization using two 4:1 MUXes and a 2:1 MUX.

takes in a 2^n-bit data input and uses an n-bit control input to get a single output. Three examples of MUXes are shown in Figure 8.4, specifically the 2:1 MUX, 4:1 MUX, and 8:1 MUX. The inputs to the MUXes are labeled I_X, the select lines (control inputs) are labeled S_X, and the output is labeled Z. Figure 8.5 shows the Karnaugh map (K-map) representation for all three MUXes. The K-map reinforces that on the basis of the select lines, the output takes on the value of one of the MUXes' inputs.

8.2.1 Implementation of larger-bit multiplexers

Similar to decoders, MUXes can be scaled up using smaller MUXes, that is, larger MUXes can be designed from smaller MUXes using a hierarchical design approach, as shown in Figure 8.6.

From Figure 8.6 and the output K-maps of Figure 8.5, we can see that the LSBs S_0 and S_1 are used to select one signal in each of the 4:1 MUXes and the MSB S_2 is used in the 2:1 MUX to select which of the two signals will be seen at the output Z. The K-map allows quick scanning of this behavior because when looked at from an MSB and an LSB point of view, the divisions are very clear: I_0, I_1, I_2, and I_3 have a chance to be selected when S_2 is 0 and I_4, I_5, I_6, and I_7 have a chance at selection when S_2 is 1. We are visually splitting the 8:1 MUX K-map into two 4:1 MUX K-maps with select lines S_1 and S_0 and then using S_2 to select between the outputs of these two 4:1 MUXes.

Over-the-counter commercial MUX chips are available, and some examples of those are the 74150 16:1 MUX, the 74151/52 8:1 MUX, the 74153 4:1 MUX, and the 74157 2:1 MUX.

8.2.2 Using multiplexers to implement boolean functions

A great use of MUXes is to select between different converging signals. A certain logic function is being performed during this selection process; hence the MUX can be used to implement Boolean functions, which is exemplified as follows

8.2.2.1 Example 1

Problem: Implement the Boolean function $L(A, B, C, D) = \Sigma(0, 2, 3, 6, 9, 10, 12, 15)$ using (a) a 16:1 MUX (b) a 8:1 MUX, and (c) a 4:1 MUX.

Solution:

(a) 16:1 MUX

First draw the K-map representation of the function L, as shown in Figure 8.7a. The K-map in Figure 8.7a is juxtaposed with that of the 16:1 MUX in Figure 8.7b. Note that the select inputs S_3, S_2, S_1, and S_0 correspond to A, B, C, and D, respectively. By visual inspection, the result displayed in Figure 8.7c can be easily obtained.

Note that we did not violate any rules here. In the 4-bit number $ABCD$, A is the MSB and D is the LSB. In $S_3 S_2 S_1 S_0$, S_3 is the MSB and S_0 is the LSB. Therefore, we were able to make a direct matching and obtain the answer.

(b) 8:1 MUX

The 8:1 MUX implementation is different from the 16:1 MUX implementation, since we need to do a bit of transformation in order to obtain our results. The first step is to draw the four-input K-map representation of L, as shown in Figure 8.8a. An 8:1 MUX takes in three select inputs, so we must eliminate one variable from the four-input K-map. We choose to eliminate the LSB, that is, D. To do this, we group the minterms that have A, B, and C in common, as shown in Figure 8.8a. For example, minterms m_0 and m_1

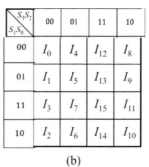

AB CD	00	01	11	10
00	0 1	4 0	12 1	8 0
01	1 0	5 0	13 0	9 1
11	3 1	7 0	15 1	11 0
10	0 1	6 1	14 0	10 1

(a)

S_3S_2 S_1S_0	00	01	11	10
00	I_0	I_4	I_{12}	I_8
01	I_1	I_5	I_{13}	I_9
11	I_3	I_7	I_{15}	I_{11}
10	I_2	I_6	I_{14}	I_{10}

(b)

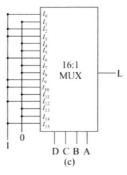

(c)

Figure 8.7 (a) K-map representation of L, (b) 16:1 MUX K-map, and (c) 16:1 MUX implementation of *L*.

have $A = 0$, $B = 0$, and $C = 0$ in common. The only change between these two minterms is D, where $D = 0$ for m_0 and $D = 1$ for m_1.

The next step is to write a three-input K-map for *L*. We do this by comparing D to the values of *L*. For the case $A = 0$, $B = 0$, and $C = 0$, when $D = 0$, $L = 1$ but when $D = 1$, $L = 0$. Essentially *L* takes on the value of $\overline{D}$. So in the three-input K-map, for $A = 0$, $B = 0$, and $C = 0$, the value of *L* is $\overline{D}$. The same process is applied to fill in the other spots on the three-input K-map presented in Figure 8.8b.

The next step is to juxtapose the three-input K-map for *L* with the 8:1 MUX K-map (redrawn in Figure 8.8c) and visually map one input to another in order to obtain the implementation shown next. *ABC* essentially corresponds to $S_2S_1S_0$, making $I_0 = \overline{D}$, and so on.

(c) 4:1 MUX

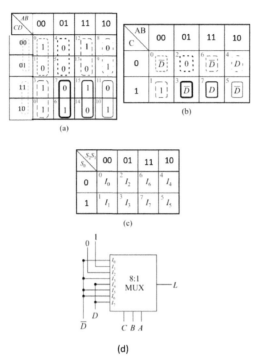

Figure 8.8 (a) Four-input K-map representation of L, (b) three-input K-map implementation of L, and (c) 8:1 MUX K-map (d) implementation of function L by 8:1 MUX.

Start solving the problem by writing down the four-input K-map representation of L (redrawn in Figure 8.9a). A 4:1 MUX takes two select inputs; therefore we need to eliminate two variables from the four-input K-map of Figure 8.9a. We choose to eliminate the lower significant bits, C and D. To do this, we group terms where A and B are the same, as shown in Figure 8.9a.

The next step involves writing of the two-input K-map representation of L. We do this by condensing the groupings of Figure 8.9a into expressions. Each grouping will correspond to one expression in the two-input K-map. Most of these can be done by inspection, for example, when $A = 0$ and $B = 0$, $L = 1$ for all cases except when $C = 0$ and $D = 1$. This one case will essentially determine our expression and can be written as $L = 1$ whenever either $C = 1$ or $D = 0$. When $A = 1$ and $B = 1$, we see that $L = 1$ whenever $C = D = 0$ or $C = D = 1$. This implies the equivalency (or XNOR) function, therefore giving the

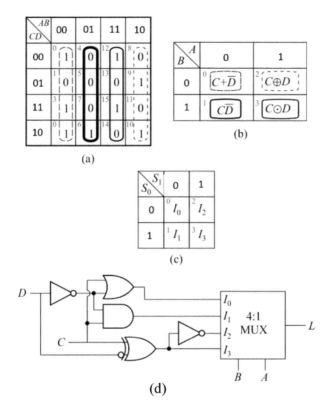

(a)

(b)

(c)

(d)

Figure 8.9   (a) Four-input K-map representation of L, (b) two-input K-map implementation of L, and (c) 4:1 MUX K-map (d) implementation of function L by a 4:1 MUX.

$A = 1$ and $B = 1$ expression of $C \odot D$. Similarly, the other expressions filled in the two-input K-map in Figure 8.9b can be obtained.

The last step is to juxtapose the 4:1 MUX K-map (redrawn in Figure 8.9c) with the two-input K-map implementation of L and visually map the select lines to the variables and the MUX inputs to the expressions. The circuit implementation next is the result of the implementation of L using a 4:1 MUX.

To reinforce the process of starting with a four-input K-map and reducing it to a two-input K-map for the implementation of a function, another example is presented.

8.2.2.2 Example 2

Problem: Implement the Boolean function $L(A, B, C, D) = \Sigma(0, 2, 3, 6, 9, 10, 12, 15)$ using a 4:1 MUX.

Solution: With A as the MSB and D as the LSB, write the four-input K-map representation of the function L, as shown in Figure 8.10a. A 4:1 MUX has two select lines, so we must reduce the four-input K-map to a two-input K-map. We do this eliminating the lower significant bits, i.e. C and D. Hence we group the terms that have A and B in common as shown in Figure 8.10a. The next step involves reducing the grouped terms into expressions in the two-input K-map. For example, for $A = B = 0$, the output $L = 1$ whenever $D = 1$; hence this grouping reduces to $L = D$ in the two-input K-map. Following the same logical thought, expressions for the other groupings can be determined to obtain the two-input K-map of L in Figure 8.10b.

The last step involves the juxtaposition of the two-input K-map implementation of L with that of the 4:1 MUX K-map (shown in Figure 8.10c). The circuit can then be drawn by visually equating variables in one K-map to another, as presented here.

With the discussion of decoders and MUXes on a theoretical level, reasonable engineering hardware descriptions will follow. As previously mentioned, decoders and MUXes can be obtained in MSI ICs. These ICs are used in the lab to implement functions, hence their introduction.

8.3 MSI Building Blocks

8.3.1 Decoders

Binary decoders, as previously mentioned, if enabled activate only one of many outputs. Two decoders will be discussed in this section, namely the 74139 dual 2:4 decoder and the 74138 3:8 decoder.

8.3.1.1 The 74139 decoder

The logic symbol of the 74139 decoder is shown in Figure 8.11, courtesy of National Semiconductor DM74LS138:DM74LS139 datasheet. The 74139 package contains two identical 2:4 decoders. Pins 1 through 7 on the package correspond to pin connections to the first decoder, while pins 9 through 15 correspond to pin connections to the second decoder. From the 74139 symbol, the number 1 before a pin name signifies that the pin belongs to the first decoder, while a 2 before the pin name signifies the pin belongs to the second

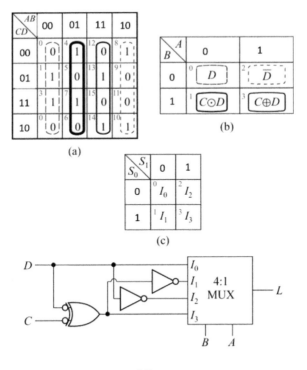

(d)

Figure 8.10   (a) Four-input K-map representation of L, (b) two-input K-map implementation of L, and (c) 4:1 MUX K-map (d) implementation of function L by a 4:1 MUX.

decoder. Y_X signify outputs, G signifies the enable inputs, and A and B signify address inputs. The address inputs (A and B) are active-HIGH inputs, the enable inputs are active-LOW, and the outputs are active-LOW.

The active-LOW and active-HIGH designations are important because this means that the 74139 decoder is enabled whenever either G_1 or G_2 is LOW. Due to the output low designation, the selected output will be 0, while the unselected outputs will be 1. This is an inverted form of the active-HIGH truth tables in Figure 8.2b. *Note: The active-HIGH and active-LOW decisions are made because inverting gates are implemented with less transistors (hence faster) than noninverting gates. The decision to use active-LOW logic is not made to confuse.*

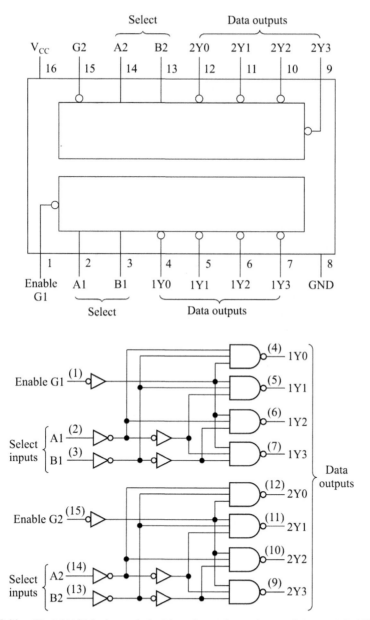

Figure 8.11 (Top) 74139 logic symbol with package pin numbers and (bottom) 74139 logic circuit implementation.

8.3.1.2 The 74138 decoder

The 74138 decoder is a 3:8 decoder in a 16- pin package. The logic symbol and schematic are shown in Figure 8.12. From the symbol, the 74138 decoder has three enable signals (i.e., G_X) that must be activated in order for the chip to perform the decoding operation. G_1 is active-HIGH, and the G_{2X} are active-LOW. The address lines are named A, B, and C, and just like the 74139 decoder, these are active-HIGH inputs. Similar to the 74139 decoder, the 74138 outputs are all active-LOW outputs. So, for example, when the address, CBA = 000, the output Y_0 will be 0, while the other outputs will be 1. C is the MSB in the 74138 decoder.

Similar to Example 2, decoder ICs can be cascaded together to build larger decoders. A cascading example is shown in Figure 8.13, where the 74138 decoder is used to make a 4:16 decoder. As in Example 2, the MSB is used to select either the first eight minterms or the last eight minterms.

EN is the enable signal, S_3 through S_0 the select signals, and D_0 through D_{15} the output signals. Here we choose to use an EN signal so that we may deactivate both decoders, if necessary. S_3, the MSB, is fed into an active-LOW enable input on the top decoder and to an active-HIGH enable input on the bottom decoder. Other than the change in the number of enable signals of the 74138 decoder, the implemented 4:16 decoder is identical to the ideal decoder presented in Example 2.

Most circuit verification in this course is done through Verilog; hence the Verilog implementation of the 74138 decoder is discussed. The first line of the code in Figure 8.14 defines the part (vlog74138) and lists the ports of this part (G_1, G_2A, G_2B, A, B, C, Y). The next two lines define which ports are inputs and which ports are outputs. Note the output Y is defined as a bus instead of individual Y_1, Y_2, Y_3, and so on. The code essentially determines the output of the decoder whenever any of the inputs changes. When the decoder is not enabled, the default value of making all outputs 1 is selected. When the decoder is enabled, then one of the outputs takes on the value of 0, while the other outputs are 1. This behavior is obtained with a case statement.

8.3.2 Tri-state buffers

Tri-state buffers (or three-state buffers) are logic devices that allow the implementation of a three-state logic. In addition to 0 and 1 states, this logic gate includes a high impedance state denoted as either "Z" or "Hi-Z". They are used mainly when multiple circuits share the same output line. The three-state buffer symbol is shown as either a buffer with an enable signal, as shown

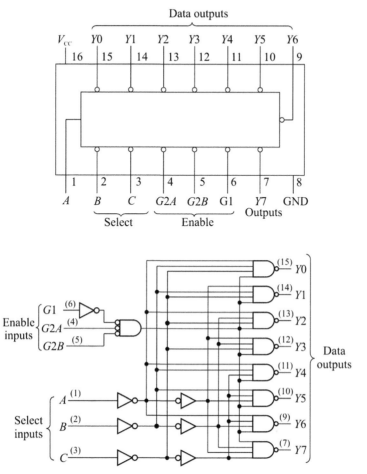

Figure 8.12 (Top) 74138 logic symbol with package pin numbers and (bottom) 74138 logic circuit implementation.

in Figure 8.15, or an inverter with an enable signal, as shown in Figure 8.16. In Figure 8.15, the buffer shown has $OUT = IN$ whenever $EN = 1$.

8.3.2.1 Application

A ubiquitous application where the tristate buffer is used is the sharing of a bus line. An example is shown in Figure 8.17. In Figure 8.17, all tristate buffer outputs are connected to the same 1-bit line. Using the 74138 decoder, only one of the tristate buffers will be activated, while the others will be in a high impedance state (i.e., "Z"). The high-impedance state can be thought of as an

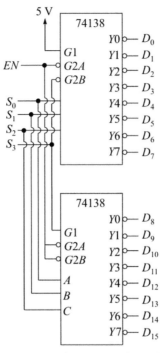

Figure 8.13   Two 74138 3:8 decoders used to make a 4:16 decoder.

```
module vlog74138 (G1, G2A, G2B, A, B, C, Y);
input G1, G2A, G2B, A, B, C; //G2A and G2B are active low
 output reg [0:7] Y;

 always @ (G1 or G2A or G2B or A or B or C)
begin
 case ({~G1, G2A, G2B, C,B,A})
0: Y=8'b01111111;
1: Y=8'b10111111;
2: Y=8'b11011111;
3: Y=8'b11101111;
4: Y=8'b11110111;
5: Y=8'b11111011;
6: Y=8'b11111101;
7: Y=8'b11111110;
default: Y=8'b11111111;
endcase
end
endmodule
```

Figure 8.14   Verilog code implementation of the 74138decoder.

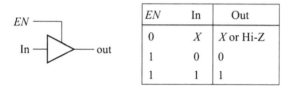

EN	In	Out
0	X	X or Hi-Z
1	0	0
1	1	1

Figure 8.15 Three-state buffer with truth table.

Figure 8.16 Multiple flavors of the three-state buffer. From left to right: active-HIGH enable buffer, active-LOW enable buffer, active-HIGH enable inverter, and active-LOW enable inverter.

open circuit. For example, if the decoder selects Q, then Q is passed to the 1-bit line, while the other buffers will behave as open circuits, disconnecting their respective inputs—M, N, O, P, R, S, and T—from the 1-bit line.

The previous example on tristate buffers presents an ideal case whereby timing glitches do not affect circuit behavior. Unfortunately, this behavior does not translate well in the real world, where noise and switching delays exist. Hence when implementing the tristate buffer, instead of using ordinary buffers, Schmitt triggers that possess hysteresis are used, as seen in the 74541 three-state buffer.

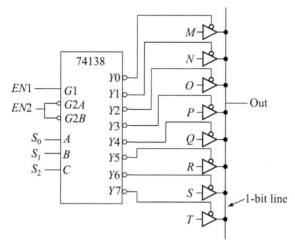

Figure 8.17 Eight sources sharing a 1-bit line.

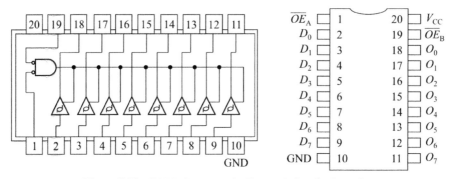

Figure 8.18   74541 three-state buffer symbol and schematic.

8.3.2.2 The 74541 three-state driver

The 74541 three-state driver has two active-LOW enable signals OE_A and OE_B, eight active-HIGH inputs D_X, and eight active-HIGH outputs O_X, as shown in Figure 8.18. The schematic shows that both enable signals OE_A and OE_B must be activated in order for the buffers to work. All buffers have the same control signal, thereby signifying that 8 bits can be controlled with these two control signals. The buffer symbols are similar to those shown in Figure 8.16 but have a parallelogram structure drawn in them to signify that the buffers have a hysteresis property to improve noise immunity.

A Verilog code implementation of the tristate buffer is shown in Figure 8.19. The code is simple enough, essentially saying whenever OE_A and OE_B are both 0, then $O = D$.

```
module vlog74541 (OEA, OEB, D, O);
input OEA,OEB; //OEA and OEB are active low
   input [0:7] D;
 output reg [0:7] O;

 always @ (OEA or OEB or D)
begin
      case ({OEA, OEB})
      0: O=D;
         default: O=8'bz;
endcase
end
endmodule
```

Figure 8.19   74541 Verilog implementation.

8.3.3 Encoders

Encoders perform the opposite function of decoders, in that they take a 2^n-bit input and convert it to an n-bit output. The simplest encoder to build is the binary encoder that takes a 2^n-bit input where only 1 bit is activated and converts it to an n-bit binary number. The equations for an 8:3 encoder are as follows:

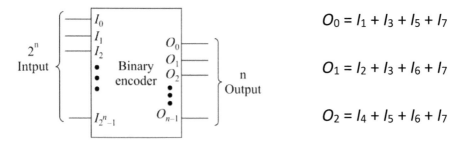

$$O_0 = I_1 + I_3 + I_5 + I_7$$

$$O_1 = I_2 + I_3 + I_6 + I_7$$

$$O_2 = I_4 + I_5 + I_6 + I_7$$

8.3.3.1 The 74148 priority encoder

Normally encoders are used in the priority encoder form. In the previous binary encoder, for the output to compute correctly, only one of the inputs must be asserted. If two or more inputs are asserted, then the result may be completely incorrect. The priority encoder avoids this problem by setting an order of primacy with respect to the inputs. Hence if two or more inputs are asserted, then the input with the highest priority is taken as being high and the others are disregarded and treated as they are not asserted. This procedure is accomplished with the 74148 priority encoder shown in Figure 8.20.

The 74148 priority encoder has the MSB I_7 as the highest priority and the LSB I_0 as the lowest priority. In Figure 8.20, the intermediate values K_X are used to make sure that only one K value is active at a time. The K values are then used to determine the A bits. The determination of the A bits is similar to that of the ordinary binary encoder previously shown.

The EI signal is the enable input signal. When EI is deactivated, that is, EI = 1, then $A_2A_1A_0$ will equal 111, GS will equal 1, and EO will equal 1. When EI is activated, then the 74148 decoder will operate with the relationships expressed in the equations in Figure 8.20. EO is the enable output signal that allows for cascading multiple priority encoders without the need for extra circuitry. GS is a signal that can be used to check if the priority encoder has received any active low inputs; hence GS is essentially a flag. If the 74148 decoder receives any active low inputs at I_X, then GS will be 0, otherwise GS

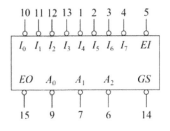

$K_7 = I_7$

$K_6 = I_6 \cdot I_7'$

$K_5 = I_5 \cdot I_6' \cdot I_7'$

...

$K_0 = I_0 \cdot I_1' \cdot I_2' \cdot I_3' \cdot I_4' \cdot I_5' \cdot I_6' \cdot I_7'$

$A_0 = K_1 + K_3 + K_5 + K_7$

$A_1 = K_2 + K_3 + K_6 + K_7$

$A_2 = K_4 + K_5 + K_6 + K_7$

$EO = \overline{I_1 \cdot I_2 \cdot I_3 \cdot I_4 \cdot I_5 \cdot I_6 \cdot I_7}$

$GS = EO'$

Figure 8.20   74148 priority encoder logic symbol and equations.

is 1. The schematic representation taken from the Signetics 74148 Encoder datasheet is presented in Figure 8.21.

8.3.3.2 Verilog implementation

The Verilog code implementation of the priority encoder is shown in Figure 8.22. The Verilog implementation of the 74148 decoder does not dive straight into equation form as might be expected. The inputs and outputs are declared with *I* declared as an 8-bit bus. All inputs and outputs have been commented to signify that they are all active-LOW. After the declaration of the inputs and outputs of the 74148 decoder, the **always** block will be accessed whenever any of the inputs (*EI* or *I*) changes.

If *EI* is 1, then the outputs $EO = 1$, $GS = 1$, and $A = 3\text{'b}111$. This takes care of the case where the priority encoder is not enabled. But when the encoder is enabled, without considering the input *I*, then *EO* should be 0 and *GS* should be 1. This case is made as the default or standard case.

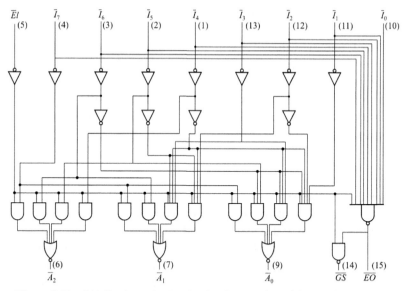

Figure 8.21 74148 schematic showing implementation of the priority encoder.

```
module vlog74148 (I, EI, EO, A, GS);
input EI; //EI is active low
input [7:0] I; //I is active low
output reg EO, GS; //EO and GS are active low
 output reg [2:0] A; //A is active low
   integer j;

 always @ (EI or I)
begin
if(~EI == 0)
       begin
         EO = 1; GS=1; A=3'b111;
       end
       else
       begin
         j=0; A=3'b111; EO=0; GS=1; //default values
         while (j < 8)
          begin
                      if(I[j] == 0)
              begin
               GS=0; EO=1; A=j;
              end
              j=j+1;
          end
       end
end
endmodule
```

Figure 8.22 74148 Verilog implementation of the priority encoder.

When any of the I inputs are 0, the decoder will execute the **if** statement. When this happens, the value of GS will be changed to 0, the value of EO will be changed to 1, and the value of A will take on the number of the input that is 0. The order of computing this loop is important. As can be seen from the code, we compute from the lowest-priority bit to the highest-priority bit. This means that if two or more inputs are 0, then the last update will always be the highest priority of all inputs and will therefore override any updates that were previously made.

8.3.4 Multiplexers

The functionality of the MUX was described in Section 8.2. Essentially the MUX is a digital switch that connects one of its multiple inputs to its single output. Like the previously discussed MSI ICs, the MUX IC has an enable signal. The enable signal is asserted separately from the MUX control inputs. For example, Figure 8.23 shows the 74151 MUX schematic and symbol. The enable signal E is an active-LOW signal, while the select signals S_2, S_1, and S_0 and the MUX inputs I_X are active-HIGH. The MUX has two outputs, Z and its complement $\overline{Z}$. The truth table for the 74151 MUX is presented in the bottom part of Figure 8.23. From the circuit schematic S_2 is the MSB and S_0 is the LSB when using the 74151 MUX IC. As with the decoders, larger MUXes, for example, the 16:1 MUX, can be built using two 8:1 MUXes and one 2:1 MUX.

Both the 74151 MUX schematic and the symbol in Figure 8.23 are obtained from the Motorola SN54/74LS151 MUX datasheet. Included in the schematic are the pin numbers for all input/output signals.

The Verilog implementation of the 74151 MUX is presented in Figure 8.24. The inputs and outputs are identified and labeled. $Z = 0$ when $E = 1$, otherwise Z gets the value of one of the inputs I_X.

8.3.5 Parity circuits

The two-input XOR gate is capable of deciphering whether both inputs are equivalent to one another. If the inputs are not the same, then the output of the XOR gate is 1, and if they are the same, the output of the XOR gate is 0. The XOR gates can be combined either in a sequence or in a tree structure (see Figure 8.25). The structures shown in Figure 8.25 are used to determine the *odd parity*, meaning that if an odd number of inputs are 1, then the output is 1. If an even number of inputs are 1, then the output is 0. To convert the

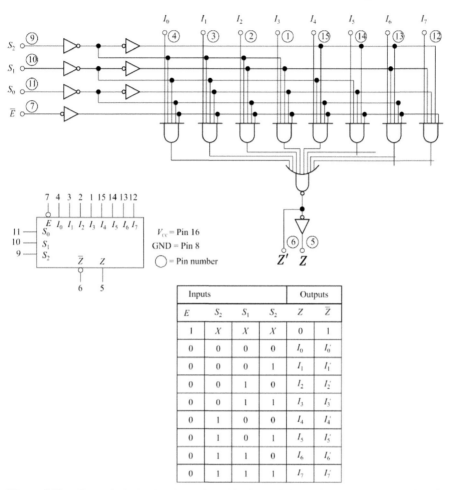

Figure 8.23 (Top) 74151 MUX schematic and logic symbol and (bottom) 74151 MUX truth table.

Inputs				Outputs	
E	S_2	S_1	S_2	Z	$\bar{Z}$
1	X	X	X	0	1
0	0	0	0	I_0	I_0'
0	0	0	1	I_1	I_1'
0	0	1	0	I_2	I_2'
0	0	1	1	I_3	I_3'
0	1	0	0	I_4	I_4'
0	1	0	1	I_5	I_5'
0	1	1	0	I_6	I_6'
0	1	1	1	I_7	I_7'

circuit to an *even-parity* circuit, the output of the odd-parity circuit should be inverted. In an MSI IC, the parity function can be realized with multiple chips, including the 74280 chip, which will not be discussed here.

```
module vlog74151 (E, S, I, Z, Zbar);
  input E; //E is active low
  input [2:0] S;
  input [7:0] I;
  output reg Z, Zbar;
  integer j=0;

  always @ (E or S or I)
  begin
  j={E, S[2], S[1], S[0]};
  if (j<8) Z=I[j];
  else Z=0;
Zbar=~Z;
  end
endmodule
```

Figure 8.24 Verilog implementation of the 74151 8:1 MUX.

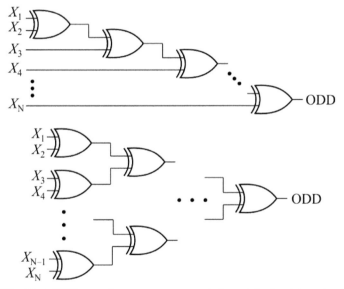

Figure 8.25 (Top) XOR cascade as a sequence or chain and (bottom) tree structure XOR cascade.

Parity circuits are very important when transmitting and receiving a code word. The parity circuit is one line of defense against using corrupt data. For example if 11110 was to be stored to memory but 11010 was received for storage, then the parity bit would be able to tell the memory controller that the received code has an error.

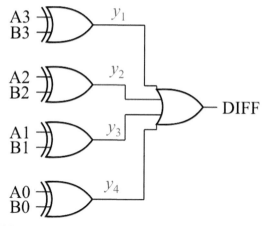

Figure 8.26   Using four XOR gates to build a 4-bit magnitude comparator.

8.3.6 Comparison circuits

8.3.6.1 Equality

Besides parity checker circuit, the XOR gate is used to build a comparison circuit, that determines if two numbers are equal to one another. A 1-bit comparator is simple in that one XOR gate will suffice to do the comparison. If the XOR gate returns a 1, then the two 1-bit numbers are not equal. This 1-bit comparator can be extended to a 4-bit comparator, as shown in Figure 8.26. Each place value of a 4-bit number A is compared against a corresponding place value of another 4-bit number B. If any place value is different, then the two numbers are not equal to one another. Figure 8.26 is its schematic representation that uses a four-input OR gate to determine if any of the XOR comparisons (y_X) returned a 1.

For an active-HIGH output, if checking that $A \neq B$, then the schematic in the true output of in Figure 8.26 is enough. If checking that $A = B$, then the output needs to be inverted, so instead of using an OR gate, a NOR gate would be used. Essentially, the magnitude comparator with active-HIGH output computes the following equation:

$$F(A=B) = \overline{A3 \oplus B3 + A2 \oplus B2 + A1 \oplus B1 + A0 \oplus B0}$$

$$F(A=B) = \overline{A3 \oplus B3} \cdot \overline{A2 \oplus B2} \cdot \overline{A1 \oplus B1} \cdot \overline{A0 \oplus B0} = (A3 \odot B3)(A2 \odot B2)(A1 \odot B1)(A0 \odot B0)$$

8.3.6.2 Greater than and less than

The process to check for if one 4-bit number is greater than another follows a systematic procedure that compares the two numbers on the basis of place values. Since each bit can be either a 1 or a 0, all exhaustive options can be listed. Given two 4-bit numbers, A and B, where A_3 and B_3 are MSBs and A_0 and B_0 are LSBs, the algorithm follows that $A>B$ under the following conditions:

1. $A_3 = 1$ AND $B_3 = 0$
2. $A_3 = B_3$ AND $A_2 = 1$ AND $B_2 = 0$
3. $A_3 = B_3$ AND $A_2 = B_2$ AND $A_1 = 1$ AND $B_1 = 0$
4. $A_3 = B_3$ AND $A_2 = B_2$ AND $A_1 = B_1$ AND $A_0 = 1$ AND $B_0 = 0$

The first condition essentially states that $A_3 > B_3$, but since we are working with binary numbers and the only options are 0 or 1, then $A_3 > B_3$ simplifies to $A_3 = 1$ and $B_3 = 0$. The same logic is used to realize the other three conditions. Starting from the largest-place-value bits, compare the bits of A and B until you find the first place value where they differ. When the bits are different, check to see if the bit value on A is larger than the bit value on B. Writing these conditions in a Boolean expression yields the following:

$$F(A > B) = A3 \cdot \overline{B3} + (A3 \odot B3) \cdot A2 \cdot \overline{B2} + (A3 \odot B3) \cdot (A2 \odot B2) \cdot$$
$$A1 \cdot \overline{B1} + (A3 \odot B3) \cdot (A2 \odot B2)(A1 \odot B1) \cdot A0 \cdot \overline{B0}$$

Each of the conditions is combined with an OR operation because if any of them are met, then $A > B$. The same procedure can be done for the $A < B$ case to obtain the following Boolean expression:

$$F(A < B) = \overline{A3} \cdot B3 + (A3 \odot B3) \cdot \overline{A2} \cdot B2 + (A3 \odot B3) \cdot (A2 \odot B2)$$
$$\cdot \overline{A1} \cdot B1 + (A3 \odot B3) \cdot (A2 \odot B2)(A1 \odot B1) \cdot \overline{A0} \cdot B0$$

8.3.6.3 The 74682 magnitude comparator IC

The equality and the greater-than operations are combined into an IC. The 74682 IC is an 8-bit comparison chip with active-HIGH inputs P and Q and active-LOW outputs $(P = Q)$ and $(P > Q)$. The 74682 schematic and logic symbol from the Motorola SN54/74LS682 8-bit magnitude comparator are provided in Figure 8.27.

The 20-pin 74682 IC can be extended to provide other comparison decisions $(P \neq Q, P \geq Q, P \leq Q, P < Q)$, as presented in Figure 8.28. The schematic presented in Figure 8.28 provides the outputs in an active-HIGH form. That is if P is equal to Q, then $P = Q$ will be 1. *Note: In this case* P = Q *from the 74682 IC will be 0 since this output is active-LOW when true, and 0 put through an inverter produces the* P = Q *we are interested in. Using a few gates, the other comparison results are determined!*

8.4 Conclusion

This chapter dealt with the theory behind decoders and MUXes and showed how to build larger-bit decoders and MUXes using smaller-bit decoders and MUXes. Afterward, implementation of Boolean functions using decoders and MUXes was shown. A very important part of circuit prototyping involves the building of circuits with over-the-counter components called MSI ICs. Some MSI circuits of decoders, MUXes, three-state buffers, encoders, and comparison circuits were discussed. In addition to these, the concept of parity was introduced.

8.5 Key Points

1. A decoder converts coded inputs into coded outputs, and the most common found is a binary decoder. Smaller binary decoders can be combined in a hierarchical manner to implement larger binary decoders.
2. Decoders provide coded outputs, and the coded outputs can be combined using logic gates to implement Boolean functions. Futhermore, outputs of binary decoders correspond to minterms or maxterms, and as such, only one of its outputs is active based on the coded input.
3. Similar to decoders, smaller multiplexers can be combined in a hierarchical manner to implement larger multiplexers.
4. Multiplexers can also be used to implement Boolean functions. If the size of the multiplexer is not too large, the K-map representations of the given Boolean function can help logic minimization by visual inspection..
5. MSI decoders, multiplexers, encoders, and tri-state buffers can be combined to implement Boolean functions. These discrete MSI chips can be purchased individually and combined on a board.

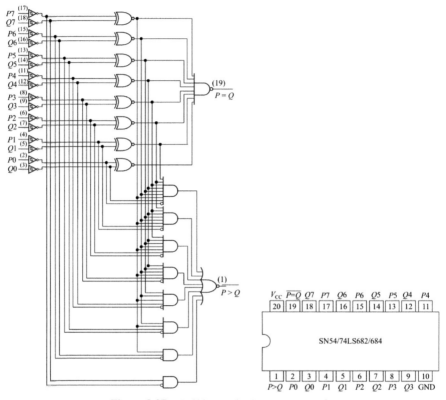

Figure 8.27   74682 magnitude comparator IC.

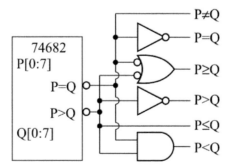

Figure 8.28   74682 IC outputs used to determine other magnitude relationships.

Sample Problems on MSI Combinational Blocks

1. Implement the following functions using only 2:4 decoders, inverters, and OR gates:

$$F1 = \sum_{X,Y,Z}(2, 4, 7)$$

$$F2 = \prod_{W,X,Y,Z}(2, 3, 5, 13, 15)$$

2. Implement the following functions using only 3:8 decoders, inverters, and OR gates:$F1 = \sum_{X,Y,Z}(1, 3, 6)$

$$F2 = \prod_{W,X,Y,Z}(0,1, 4, 7, 10,12, 14)$$

3. Generate a 16:1 MUX using 4:1 MUXes.
4. Implement $f = abc' + ab'c$ using only a 4:1 MUX. You can assume both true and complement forms of the inputs are available.
5. One type of programmable logic device uses a unit cell composed of MUXes instead of the traditional AND/OR logic architecture. A unit cell has three two-way, 1-bit MUXes and an OR gate. A specific example of a programmed cell is shown below. What is the logic function generated by this cell?

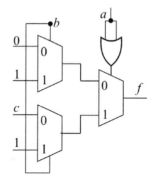

A. $z = ab + \bar{c}$ B. $z = a + bc$ C. $z = b + ac$
D. $z = ac + \bar{a}b$ E. None of the above
6. Implement the following functions using 8:1 MUXes.

   (a) $f(a, b, c) = \sum(0, 1, 3, 5)$
   (b) $f(a, b, c, d) = \sum(2, 4, 7, 11, 13, 14)$
   (c) $f(a, b, c, d) = \prod(0, 2, 5, 7, 9, 12, 15)$

7. Implement the following functions using 16:1 MUXes.

(a) $f(a, b, c) = \sum(2, 4, 7)$
(b) $f(a, b, c, d) = \sum(1, 5, 8, 9, 12, 15)$
(c) $f(a, b, c, d) = \prod(1, 4, 7, 11, 12, 14)$

8. Identify the MUX circuit that generates the function $z = ab + bc + a'b'$.

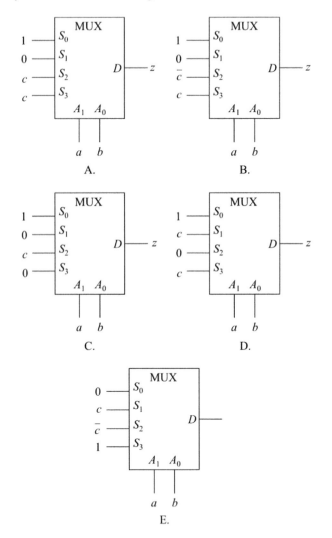

9. A barrel shifter is a combinational logic that uses banks of multiplexers as shown in the following diagram to shift an input data word by a specified number of bits. Write a Verilog program for the given barrel

shifter and show the simulation results along with your Verilog code. Note that there are three shifting possibilities in the given circuit:

- *shift(0)* is the bit0 of the shift amount. It weights $2^0 = 1$. Setting that bit shift *at least* 1 bit.
- *shift(1)* is the bit1 of the shift amount. It weights $2^1 = 2$. Setting that bit shift *at least* 2 bits.
- *shift(2)* is the bit2 of the shift amount. It weights $2^2 = 4$. Setting that bit shift *at least* 4 bits.

When shifting, the incoming bits must be set to zero for logical shifts. For example, 01001101 shifted left by 2 becomes 00110100.

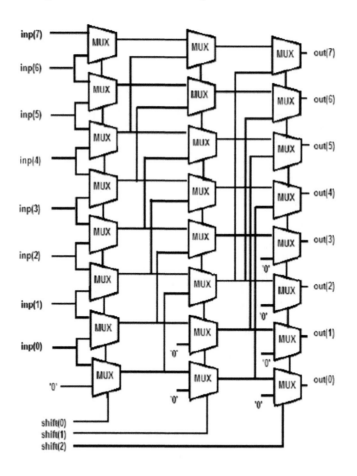

10. Design a circuit that compares two 5-bit numbers, A and B, and determines if A is greater than B. You do not need to provide the circuit schematic. You should only provide the Boolean expressions.

11. Write a Verilog program for the comparator for IC 74682 shown in Figure 8.27 and the various results of magnitude comparison test as shown in Figure 8.28. You must show the simulation results along with your Verilog code.

12. Shown below is a high-speed parity generator/checker that accepts nine bits of input data and detects whether an even or an odd number of these inputs is HIGH. If an even number of inputs is HIGH, the Sum Even output is HIGH. If an odd number is HIGH, the Sum Even output is LOW. The Sum Odd output is the complement of the Sum Even output. Write Verilog program and show the simulation results.

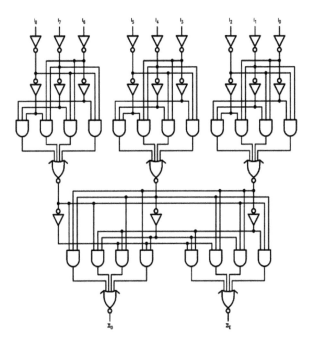

13. Write Verilog programs for the given BCD to seven-segment decoder as shown in the figure below. You must produce the simulation results at the maser and slave outputs along with your Verilog code testing the decoder for possible ten numerals.

'46A, '47A, 'L46, 'L47, 'LS47

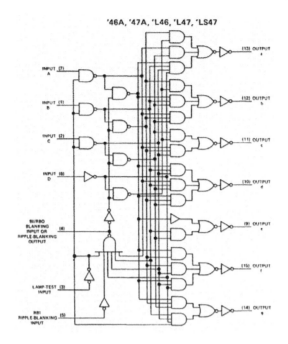

14. Write a Verilog program for a 32:1 Mux using the building block of 8:1 multiplexer shown in Figure 8.23 along with minimal additional components. You must show the simulation results along with your Verilog code.

15. Write a Verilog program for the priority encoder for IC 74148 as shown in Figure 8.21 You must show the simulation results along with your Verilog code.

16. Using one 4:1 MUX and additional gates, if needed, design a logic network that indicates an output (Z) equal to 1 if its four 8421-coded binary-coded decimal (BCD) inputs (A, B, C, D) represent a valid code word having a decimal value less than 6. The network also sets the output equal to 1 if the number of 1s in an invalid code word is more than 2. All other input patterns, otherwise, generate an output equal to 0. The select inputs of the MUX are connected to the inputs A and B, and the output of the MUX represents the output (Z) of the logic network. Solve the problem by neatly drawing a K-map (do not write a separate truth table), representing the output function Z (A,B,C,D) on the map. Write the equations for the four data inputs of the MUX. Draw a neat diagram showing all components of the logic network.

Foundations of Sequential Design: Part I

This chapter introduces flip-flops (FFs) as the fundamental elements of sequential logic design along with implementation techniques of finite state machines. At first, basic configurations of sequential circuits will be discussed, and then the flip-flops will be described.

Terms introduced in this chapter: Moore machine, Mealy machine, S-R, T, J-K, and D flip-flops

Competency Objectives: At the end of this chapter, you will be able to:

1. Explain the operation of various latches and flip-flops.
2. Make a table and derive the characteristic equation for such latches and flip-flops. State any necessary restrictions on the input signals.
3. Draw the state diagram associated with each type of flip-flop.

9.1 Taxonomy of Sequential Models

Sequential logic is defined as digital systems where the outputs do not just depend on the present values of the inputs but also on the past values of the inputs. Therefore, some storage devices must be incorporated in the circuitry in order to remember the past values of the inputs. Sequential circuits can be broadly divided into two categories: The Mealy machine and The Moore machine.

9.1.1 The mealy machine

In the Mealy machine, the outputs of the system can change when either the internal state changes or the inputs change. Figure 9.1 shows a circuit

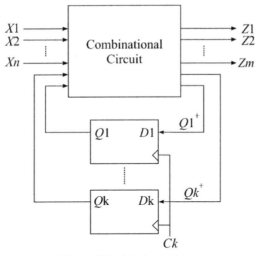

Figure 9.1   Mealy machine.

diagram of the Mealy machine, which receives both primary inputs (X_1, X_2,..., X_n) and state inputs, where the states are the outputs from D FFs (Q_1, Q_2,..., Q_k). These two distinct types of inputs are provided to the combinational circuit block, which determines the values of the primary outputs (Z_1, Z_2,..., Z_m) as well as the data inputs to D FFs. The outputs of the Mealy machine depend on the combinational circuit, and they may change asynchronously if inputs change as well as they may change synchronously with the update of FF outputs along with the system clock (Ck).

The Mealy machine generally requires less logic gates and fewer FFs in comparison to the Moore machine. However, the Mealy machine can introduce unnecessary glitches in the output waveform, since the Mealy machine output is allowed to change with the changes in primary inputs. When hardware cost is paramount for design optimization, in large digital systems the Mealy machine is preferred for its compactness. The hardware savings of a Mealy machine was attractive when the transistor count on a chip was limited to ~100,000. Currently, a billion transistors can fit in the same chip area, making the cost of a few additional transistors negligible. Hence, reliability is significantly more important than area, making the Moore finite state machine more preferable. However, in emerging technologies where the scale of integration is nowhere compared to modern CMOS technology, Mealy machine design may still be warranted

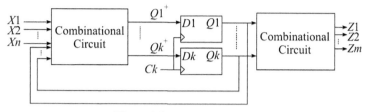

Figure 9.2 Moore machine.

in many applications. Therefore, Mealy machine analysis and synthesis will be covered in the book.

9.1.2 The moore machine

In the Moore machine as shown in Figure 9.2, the outputs of the system can only change when the outputs of the FFs change, since outputs are determined from a combinational block whose inputs are all from FFs. The outputs of the FFs are fed back to a combinational circuit that performs computations with the internal states and inputs. The output changes with each clock pulse. The next state of the system depends on the present inputs and the current state (FF outputs).

In real systems, clocked (synchronous) circuits experience various issues. Clock routing in chips is complex and clocks are power hungry, and clock skew, which will be explained later, can cause incorrect functionality. Asynchronous circuits are gathering more attention, since they do not experience these issues; but these circuits will not be discussed in this chapter.

9.2 Flip-Flops

In this section we explain the operation of FFs (the storage element) and their applications. We will describe different types of FFs and reveal characteristic tables for each type. State diagrams are introduced to visualize transitions between FF states and their underlying conditions.

9.2.1 Operation of flip-flops and their applications

A flip-flop (FF) is a memory or storage device that can store 1 bit of information (i.e., it stores 0 or 1) indefinitely, even after its inputs change. Typically, an FF has a pair of complementary outputs (shown in Figure 9.3) and one or more inputs that can cause the output to change.

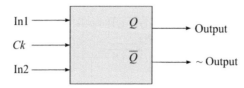

Figure 9.3   Flip-flop. In_1 and In_2 are inputs. C_k is the clock input to the flip-flop. It synchronizes the changes in the outputs Q and $\overline{Q}$.

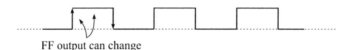

FF output can change

Figure 9.4   Clock signal. The FF output can change on the edge and during the time period the clock is high.

Output Q and its complement $\overline{Q}$ are held constant even after the FF inputs are changed. Outputs change only when the clock changes, that is, outputs are synchronized with the clock. For FFs the output can change on several occasions, such as at the edge of the clock or during the time period when the clock is at a specific level (i.e., logic level high or low), as shown in Figure 9.4.

9.2.2 Classification of flip-flops

Principally, there are four types of FFs:

- Set-Reset (S-R) FF
- Toggle (T) FF
- Generalized (J-K)FF
- Data or Delay (D) FF

D FFs are the most common type storage devices used in sequential circuits. T or S-R FFs can be used to implement counters. S-R FFs can be configured to implement T FFs and J-K FFs. The next subsections will discuss these four types of FFs in detail. Initially we will consider clock-less or un-clocked flip-flops that act as basic storage or memory elements.

9.2.2.1 S-R flip-flop
9.2.2.1.1 Characteristic table
Figure 9.5 shows an S-R FF symbol, and Figure 9.6 shows the input-output characteristics. The present state of the FF is reflected as Q, which also serves as the output of the FF. As seen in the characteristic table, the next state, $Q+$

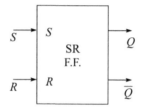

Figure 9.5 S-R flip-flop symbol.

S	R	Q+	Operation
0	0	Q	HOLD
0	1	0	RESET
1	0	1	SET
1	1	?	Invalid

Figure 9.6 S-R flip-flop characteristic table. $Q+$ is the next state of Q. If $Q(t)$ is the value of Q at time t, then $Q+$ is $Q(t + 1)$.

of the S-R flip-flop does not change when both S and R inputs are 0. This input combination yields the HOLD operation, meaning the FF is holding its current state, Q. When S is 0 and R is 1, the next state becomes 0. This input combination results in the RESET operation. When S is 1 and R is 0, the SET operation is performed and the next state becomes 1. The last row in the characteristic table indicates that both S and R cannot be 1 simultaneously. This operation is **invalid** for reasons that will be apparent when we discuss the gate implementation of the S-R FF.

With the characteristic table defined, we will define the next state, $Q+$, in terms of inputs, S and R, and present state, Q. The relationship shown in the characteristic table is represented in the Karnaugh (K) map in Figure 9.7. In the K-map, the invalid next states are represented by "don't cares" or X's in order to obtain *a compact design of the S-R flip-flop*.

From the K-map in Figure 9.7, the next state can be written as a sum of products as $Q^+ = \overline{R}Q + S$. This relationship is termed the *characteristic equation* of an S-R FF. The equation indicates that we can relate the next state, $Q+$, to the present state, Q. Note that by including the X's in the expression for Q^+, the next state will be 1 ($Q^+=1$) when $S = R = 1$, no matter what the present state is (i.e., $Q = X$)

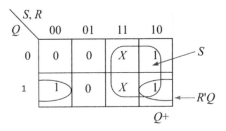

Figure 9.7   The Karnaugh map for an S-R flip-flop.

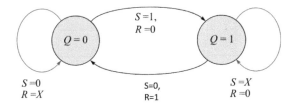

Figure 9.8   S-R flip-flop state diagram. Arrows represent state transitions. The conditions for the transitions are listed next to the arrows.

9.2.2.1.2 State diagram

Figure 9.8 shows the two Boolean states for Q, 0 and 1. Inputs S and R control the transition between these two states. If the current state is 0 (i.e., $Q = 0$) and the input S is 0, there can be two operations possible for an S-R FF, depending on the value of R. If R is 0, the FF is supposed to HOLD its state; thus the next state will be 0 (i.e., $Q^+ = 0$). If R is 1, the FF performs RESET operation, setting the next state to 0. Therefore as long as S is 0 and Q is 0, the current state will remain unchanged (i.e., $Q = Q^+ = 0$).

If S is 1 and R is 0, then the FF performs the SET operation, setting the next state to 1 (i.e., $Q^+ = 1$). If the current state is 1 and the input R is 0, there can be two possible operations for the S-R FF. If S is 0, the FF is in HOLD its state; therefore, next state stays at 1. If S is 1, the FF performs the SET operation, making the next state still remain 1. Therefore the state of the FF will not change as long as R is 0 and Q is 1 (i.e., $Q = Q^+ = 1$). If R is 1 and S is 0, the FF performs the RESET operation and the next state becomes 0. The input combination $S = 1$ and $R = 1$ is not permitted; therefore, it is not included in the state diagram.

The S-R FF is a fundamental storage device that can be deployed to implement other types of FFs, as explained in the following sections.

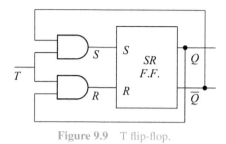

Figure 9.9   T flip-flop.

T	R	S	Q+	Operation
0	0	0	Q	HOLD
1	Q'	Q'	Q'	TOGGLE

Figure 9.10   T flip-flop characteristic table.

9.2.2.2 T flip-flop

A T FF has a single input, designated as T, and can be obtained from an S-R FF by letting the input $R = Q \cdot T$ and input $S = \overline{Q} \cdot T$. Figure 9.9 shows a schematic of the T FF. Two two-input AND gates are introduced to generate required S and R inputs of the S-R FF.

9.2.2.2.1 Characteristic table

The characteristic table of the T FF is shown in Figure 9.10, and the conceptual implementation based on the S-R FF is presented in Figure 9.9. If the T input is 0, R and S become 0. In this case, if we refer back to the S-R FF characteristic table, we see that the FF is performing a HOLD operation. Therefore when $T = 0$, the T FF performs the HOLD operation. If T is 1, then $R = Q \cdot T$ becomes $R = Q$, and $S = \overline{Q} \cdot T$ becomes $S = \overline{Q}$. In this case (i.e., $T = 1$), if Q was initially 0, then R is 0 and S is 1; thus the S-R FF performs the SET operation, setting the next state to 1 ($Q^+ = \overline{Q} = 1$). This indicates that the output of the T FF **toggle sits output** from 0 to 1. If Q was initially 1, then R is 1 and S is 0, and the S-R FF performs the RESET operation, setting the next state to 0 (i.e., $Q^+ = \overline{Q} = 0$). Therefore, the output of the T FF toggles from 1 to 0. Essentially, the $T = 1$ case is the TOGGLE operation, complementing or toggling the output value. The K-map representation (Figure 9.11) can be used to determine Q^+, given T and Q.

T	0	1
Q		
0	0	1
1	1	0

$Q+$

Figure 9.11   Karnaugh map for a T flip-flop.

From the K-map in Figure 9.11, $Q+ = T \cdot \overline{Q} + Q \cdot \overline{T} = T \oplus Q$, which is the characteristic equation of the T FF. The T FF state diagram (Figure 9.12) shows the transition conditions between the two states, $Q = 0$ and $Q = 1$. The FF changes state when T is 1 and stays at the same state when T is 0.

9.2.2.2.2 State diagram

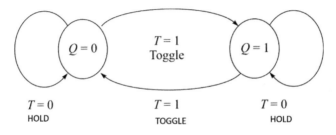

Figure 9.12   T flip-flop state diagram. Arrows represent state transitions. The conditions for the transitions are listed next to the arrows.

9.2.2.3 J-K flip-flop[1]

S-R FFs have a fundamental problem. The inputs are not allowed to be 1 simultaneously; thus, it becomes the responsibility of the circuit designer to make sure this never occurs. To account for the case when both inputs are 1, we design the J-K FF (Figure 9.13). The circuitry we use for a J-K FF is the same as that used for a T FF. In the J-K FF case, we introduce two inputs, designated as J and K, to set the node $R = Q \cdot K$ and the node $S = \overline{Q} \cdot J$.

---

[1]J-K Flip-Flop was purportedly named after Jack Kilby of Texas Instruments who received a Nobel Prize in Physics (2000) for inventing monolithic integrated circuits (ICs) that revolutionized electronics manufacturing now containing few billions of transistors in a single IC.

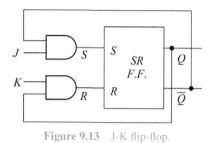

Figure 9.13 J-K flip-flop.

9.2.2.3.1 Characteristic table

The J-K FF is a superset of the S-R FF. The first three rows of both J-K (Figure 9.14) and S-R (Figure 9.6) FF characteristic tables are identical. If both inputs, J and K are 0, then the FF performs the HOLD operation. If J is 0 and K is 1, then the FF performs the RESET operation, making Q^+ equal to 0. If J is 1 and K is 0, the SET operation is performed and the next state, Q^+ becomes 1. For these input pairs, the J-K FF operates just like the S-R FF. The difference between both FFs is that the J-K FF's inputs are allowed to be 1 simultaneously. If the inputs J and K are both 1, the FF performs the TOGGLE operation. This functionality is the same as the T FF's toggle operation. Therefore, a J-K FF contains the features of both S-R and T FFs.

To determine the next state in terms of inputs and the current state, we draw the K-map (Figure 9.15). We can draw two groups of two minterms.

J	K	Q^+	Operation
0	0	Q	No Change [S=R=0]
0	1	0	RESET [S=0, R=Q(t)]
1	0	1	SET [S=Q'(t),R=0]
1	1	Q'	TOGGLE [S=Q' and R=Q]

Figure 9.14 J-K flip-flop characteristic table.

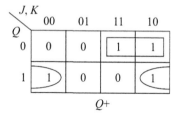

Figure 9.15 Karnaugh map for a J-K flip-flop.

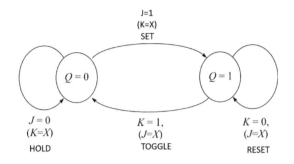

Figure 9.16   J-K flip-flop state diagram. Arrows represent state transitions. The conditions for the transitions are listed next to the arrows.

The result is $Q+ = J \cdot \overline{Q} + Q \cdot \overline{K}$. This is the characteristic equation of a J-K flip-flop. A T FF can be obtained from a J-K FF by setting both inputs to T of T FF (i.e., $J = K = T$).

9.2.2.3.2 State diagram

The state diagram (Figure 9.16) indicates that transition from $Q = 0$ to $Q = 1$ occurs if the SET operation or the TOGGLE operation is performed, meaning if $J = 1$ and $K = 0$ or if $J = 1$ and $K = 1$. Hence, from the $Q = 0$ state, regardless of the value of K, Q will change from 0 to 1 if J is 1. Transition from state $Q = 1$ to $Q = 0$ occurs if the RESET operation or the TOGGLE operation is performed, meaning if $J = 0$ and $K = 1$ or if $J = 1$ and $K = 1$. From a $Q = 1$ state, irrespective of the value of J, Q will change from 1 to 0 if K is 1. If the initial state is 0, no state transition will occur if the RESET or the HOLD operation is performed. RESET occurs when $J = 0$ and $K = 1$, while HOLD occurs when $J = 0$ and $K = 0$. Thus, when starting from an initial state of 0, the state will not change as long as J is 0. If the initial state is 1, no state transition will occur if the SET or the HOLD operation is performed. The SET operation inputs are $J = 1$ and $K = 0$, while HOLD inputs are $J = 0$ and $K = 0$. Thus, if starting from the initial state 1, the state will not change as long as K is 0.

9.2.2.4 D flip-flop

D FF is used to insert delay in the digital systems, hence the name D (delay) FF. The flip-flop is generally utilized for state machine implementation. A D FF can be obtained from a J-K or S-R FF by adding an inverter, as shown in Figure 9.17. Therefore, when a D FF is converted from a S-R FF, the values

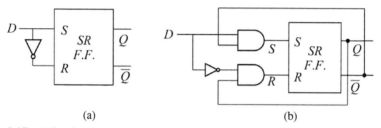

(a)   (b)

Figure 9.17   D flip-flop. (a) S-R FF configured as a D FF and (b) S-R FF configured as a J-K FF, which is then configured as a D FF.

of S and R are derived as $S = D$, $R = D'$ as shown in Figure 9.17a. If the D FF is derived from J-K FF, then $J = D$ and $K = D'$ as illustrated in Figure 9.17b where $S = DQ'$ and $R = D'Q$.

Figure 9.18 shows the characteristic table of a D FF. D FFs can only perform SET and RESET operations, depending on the input state. If the input is 0, the next state is 0, and if the input is 1, then the next state is 1. The characteristic equation of a D FF is straightforward and can be directly seen from the characteristic table. Therefore, $Q^+ = D$.

When we compare the characteristic tables of D and J-K FFs, we see that if $J = \overline{K}$, then $Q^+ = D$. Therefore, as discussed before, a J-K FF can be converted into a D FF as shown in Figure 9.17b. The same thing can be achieved with an S-R FF by connecting the complement of the S input to the R input as shown in Figure 9.17a.

The D FF state diagram is shown in Figure 9.19. If the input D is 1, the next state will be SET to $Q^+=1$ no matter what the initial state is. If the input D is 0, the next state will be RESET to $Q^+ = 0$. Therefore, the D FF's next state does not depend on the current state.

D	Q^+	Operation
0	0	RESET
1	1	SET

Figure 9.18   D flip-flop characteristic table.

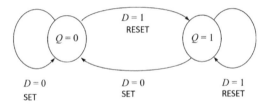

Figure 9.19   D flip-flop state diagram. Arrows represent state transitions. The conditions for the transitions are listed next to the arrows.

9.3 Conclusion and Key Points

1. There are two types of sequential logic: synchronous logic that is driven by clock pulses and asynchronous logic that changes its output and internal states when the inputs change (this is known as fundamental mode of operation).

2. There are two models for synchronous logic design: Moore machine and Mealy machine. In case of Moore machine, the output signals change only with the clock pulses, i.e., the output signals remain constant in each present state of the machine. In case of Mealy machine, the output is a function of the present inputs as well as the present states.

3. Because output signals do not remain constant in the present state, they may change with the change of input signals before the machine goes to the next state. Therefore, Mealy machines may inadvertently produce false signals or glitches due to design error or variation in signal timing from one chip to another.

4. There are primarily two types of flip-flops: S-R and J-K flip-flops. Both T and D flip-flops are derived from J-K flip-flops.

5. S-R flip-flop produces Invalid output where $Q = Q'$ when $S = R = 1$, i.e., both Set (S) and Reset (R) inputs are asserted by setting them High, simultaneously. If $S = R = 0$, then S-R flip-flop holds its Q and Q' values correctly. When $S = 1, R = 0$, the S-R flip-flop is in Set mode of operation and the output, $Q = 1$ and $Q' = 0$. On the other hand, when $S = 0, R = 1$, the S-R flip-flop in Reset mode of operation and the output, $Q = 0$ and $Q' = 1$.

6. J-K flip-flop overcomes the limitation of S-R flip-flop by introducing Toggle operation when $J = K = 1$, and the output of the flip-flop inverts, i.e., the next state of the flip-flop is given by: $Q^+ = Q'$ and $(Q^+)' = Q$.

7. Note that J-K flip-flop can be obtained from an S-R flip-flop by setting $S = JQ'$ and $R = KQ$.

8. Note that a D flip-flop can be obtained from a J-K flip-flop by setting $J = K' = D$.
9. Note that a T flip-flop can be obtained from a J-K flip-flop by setting $J = K = T$.

Problems on Flip-Flops

1. What is the difference between Moore and Mealy machines? For the system described below, what type of machine-based implementation would be the best? Discuss your reasoning behind your answers.

   (a) A controller where the controller needs to stop asserting a signal as soon as an input is detected
   (b) A controller that has a noisy input where the input glitches before settling to its final value

2. An MNFF is defined by the symbol and function table given below. A finite-state machine is to be built using MN FFs. In the design it is necessary to produce the state transition $(Q = 1)$ $(Q^+ = 0)$. Which input combination will cause this input ("X" means "don't care")?

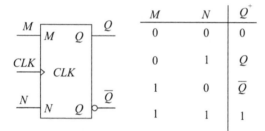

M	N	Q^+
0	0	0
0	1	Q
1	0	$\overline{Q}$
1	1	1

   A. $M = X, N = 1$
   B. $M = 1, N = X$
   C. $M = 0, N = X$
   D. $M = X, N = 0$
   E. None of the above

3. Using a minimum number of gates implement a DFF using a TFF. Show the circuit diagram and the equations relating D and T.
4. Using a minimum number of gates implement a TFF using a J-KFF. Show the circuit diagram and the equations relating J, K, and T.
5. Implement a T FF using only a D FF and a 2:1 multiplexer.
6. For the circuits shown below, which of the following statements are true?

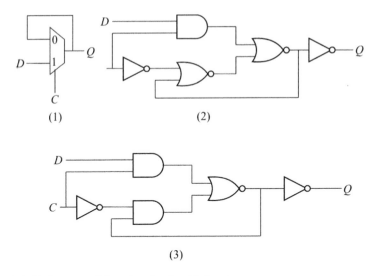

(1)   (2)

(3)

(a) Only one of the circuits is a D latch.
(b) Only circuits 1 and 2 are D latches.
(c) Only circuits 2 and 3 are D latches.
(d) Only circuits 1 and 3 are D latches.
(e) All three circuits are D latches.

7. Say you have a summer job at the famous Disney Electronics Laboratory, where bright EECS students are regularly recruited to invent digital gizmos. Your supervisor asks you to design a two-input; single-output FF (unclocked) using only one 2:1 multiplexer. The inputs to the FFs are A and B, and the output of the FF is Q. Assume A is connected to the data input I_0 (not to I_1) of the 2:1 multiplexer. If the present state of the FF is 1 and the next state is 0, then which of the following inputs are true?

(i) $A = 0, B = 0$
(ii) $A = 1, B = 0$
(iii) $A = 0, B = 1$
(iv) $A = 1, B = 1$ (B is connected to a select input.)

A. Only (iv) is true.
B. Only (i) is true.
C. Both (i) and (ii) are true.
D. Both (iii) and (iv) are true.
E. None of the above is true.

8. An FF that has inputs U and V is constructed from an edge-triggered SR FF (same behavior as an SR latch, but changes state on the rising clock edge) and a couple of gates.

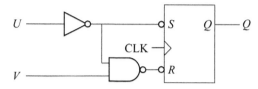

What is the characteristic function of the FF?

A. $Q^+ = V + UQ$
B. $Q^+ = U + V'Q$
C. $Q^+ = VQ + UQ'$
D. $Q^+ = VUQ + U + V'$
E. $Q^+ = V'U + Q'$

9. An FF that has inputs X and Y is constructed from an edge-triggered JK FF and a NORgate. Determine the characteristic function of the FF.

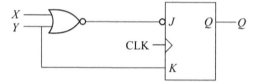

10. We can convert a D FF to a T FF using the circuit shown above by adding a logic gate G, as shown. The gate should be:

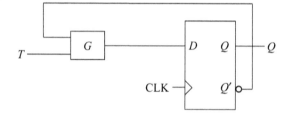

A. A NOR
B. An XOR
C. A NAND
D. An OR
E. An XNOR

Foundations of Sequential Design: Part II

This chapter presents the gate implementation of unclocked flip-flops and discusses their operation using timing diagrams. After the buildup of flip-flop circuitry, the clock signal is introduced to add the property of synchronization. Problems associated with clocked implementations are discussed, and several examples that portray flip-flop circuit analysis are presented.

Terms introduced in this chapter: gate implementation, clocked flip-flop, and register file.

Competency Objectives: At the end of this chapter, you will be able to:

1. Design the gate-level structure of an S-R flip-flop in NOR-NOR configuration from the product-of-sums characteristic equation of the flip-flop. Design the gate-level structure of an S-R flip-flop using NAND-NAND configuration from the sum-of-products characteristic equation of the flip-flop.
2. Recognize that S-R flip-flop not only provides invalid output pair, Q = Q' when S = R = 1, it breaks into oscillation if after asserting both S and R inputs High, they are simultaneously asserted Low, i.e. in the Hold mode. Note that both Q and Q' outputs oscillate and Q = Q'.

3. Draw a timing diagram relating to the input and output of various types of latches and flip-flops.
4. Show how latches and flip-flops can be constructed using gates. Analyze the operation of a flip-flop that is constructed of gates and latches.

10.1 Gate Implementation of Flip-Flops and Timing Diagrams

This section discusses basic data storage units and analyzes their essential functions. The organization in this section flows from rudimentary to more advanced storage units. Rudimentary storage units are composed of simple inverters, while more advanced storage units with increased functionality are composed of complex gates. By incorporating complex gates, the advanced storage units are able to perform operations such as SET, RESET, and HOLD, which were introduced in Chapter 9.

10.1.1 Simple inverters without feedback

In a simple chain of inverters, like in Figure 10.1, the output of the first inverter, Q, serves as the input to the second inverter. In this configuration, the output of the first inverter, Q, is equal to $\overline{X}$, where X is the input of the first inverter. The output of the second inverter is $\overline{Q}$, which is equal to X. This example provides a basis for the types of circuits encountered so far, where data flows from input to output, setting intermediate nodes as the data signal propagates through the chain. $\overline{Q}$ and Q are directly controlled by X, so previous values of Q and $\overline{Q}$ are unimportant in analyzing this circuit.

The next section will introduce a circuit whereby the previous values on a wire connection matter.

10.1.2 Single inverter with feedback (oscillator)

The simplest circuit to show the effect of memory is a single inverter with feedback, as depicted in Figure 10.2a. If the input, X, and the output, Q, of an inverter are connected, oscillating waveforms are observed, as shown

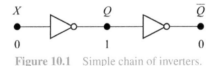

Figure 10.1   Simple chain of inverters.

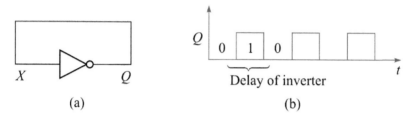

(a)   (b)

Figure 10.2 (a) Oscillator realization with a single inverter in feedback configuration and (b) output waveform.

in Figure 10.2b. To verify this behavior, assume that the output of the inverter is initially 0. Since the inverter's output is connected directly to its input, this assumption also sets X to 0. If X is 0, then Q switches to 1 after a time delay equal to the delay of the inverter. Recall that logic gates have a finite propagation delay and do not provide outputs instantaneously. After Q (and X) switch to 1, the inverter begins the NOT operation over again, causing Q (and X) to switch back to 0 after an inverter gate delay. Hence the output of the inverter oscillates between 0 and 1 with a period equal to twice the delay of the inverter[1].

10.1.3 Two inverters with feedback (flip-flop)

If one inverter in feedback makes an oscillator, then what do two inverters in a feedback configuration make? Figure 10.3 shows a two-inverter feedback configuration and depicts graphically two possible states, storing 1 or storing 0. If the input of the first inverter is 0, then its output Q is 1. If the input of the second inverter is 1, then its output $\overline{Q}$ is 0. Since the output of the second inverter is also connected to the input of the first inverter and they both have the same value, there will be no change propagated to Q. This configuration is stable, meaning the outputs of the inverters do not oscillate. The same stability

---

[1]This illustration of oscillatory behavior with a single (odd) inverter was described in Chapter 4 with a chain of three (odd) of inverters connected in the form of a ring. In practice, with a single inverter as in Figure 10.2, a crystal is incorporated in the feedback path to build an accurate clock oscillator for which the time period depends on the crystal frequency. Without any delay element in the feedback path, a single inverter may not oscillate in some cases and both the output and the input of the inverter may be held at about 0.5 V_{DD}. This is known as metastability and is different from the two stable states of a flip-flop, pertaining to the output is 0 (typically, Ground or V_{SS}) or 1 (typically, V_{DD}). A flip-flop may also enter into the metastability state due to timing violation of the input signals to the flip-flop. This is described in Section 10.2 in this chapter.

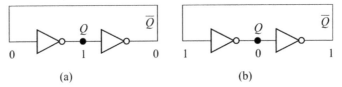

(a)   (b)

Figure 10.3   Two-inverter chain (a) storing 1 and (b) storing 0.

is observed if the input of the first inverter was 1. The information, logic 1 or 0 stored as the state of Q, will stay the same indefinitely. This setup is a rudimentary way of obtaining a storage unit.

10.1.4 S-R latch

From basic principles, the two-inverter storage does not provide the flexibility necessary to manipulate data stored in the inverter chain. Two extra inputs are introduced to convert the inverter chain into a set-reset (S-R) latch, as depicted in Figure 10.4a. Two-input NOR gates are used instead of inverters, since inverters only have one input. Analyzing the NOR implementation provides the truth table shown in Figure 10.4b. The table affirms that the S-R latch is capable of storing states 0 and 1 when performing the HOLD operation and the latch can set Q to either 0 or 1 when performing the RESET or SET operation, respectively.

The S-R latch's HOLD operation promotes the concept that the latch exhibits behavior that is dependent on its previous state. This realization introduces the concept of current state Q and the next state $Q+$. The S-R latch expression for the next state $Q+$ can be obtained by drawing the Karnaugh maps (K-maps) in Figure 10.5. There are two ways to obtain expressions using the K-map: either group the minterms to obtain a sum of

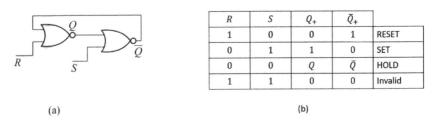

R	S	Q_+	$\bar{Q}_+$	
1	0	0	1	RESET
0	1	1	0	SET
0	0	Q	$\bar{Q}$	HOLD
1	1	0	0	Invalid

(a)   (b)

Figure 10.4   S-R latch. (a) Schematic showing inputs and outputs and (b) truth table showing different operations.

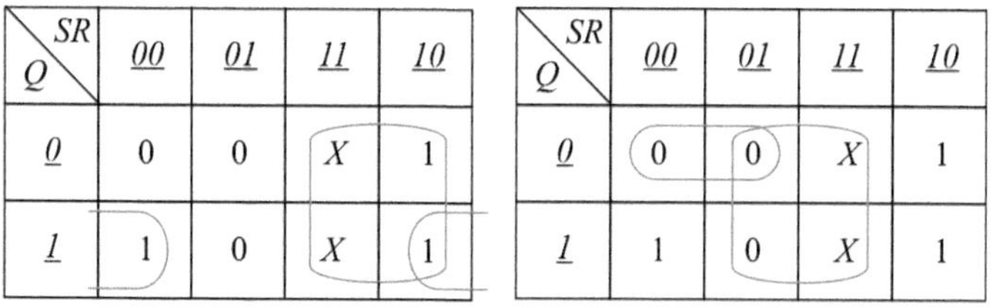

SR Q	00	01	11	10
0	0	0	X	1
1	1	0	X	1

(a)

SR Q	00	01	11	10
0	0	0	X	1
1	1	0	X	1

(b)

Figure 10.5 K-map groupings for S-R latch: (a) minterm groupings and (b) maxterm groupings.

products (SOP) expression or group the max terms to obtain a product of sums (POS) expression. An SOP expression can be used to obtain NAND gate-based implementation, while a POS expression can be used to obtain a NOR gate-based implementation.

Figure 10.5a shows 1's grouped to obtain the SOP expression $Q+ = S + \overline{R}Q$. Figure 10.5b shows 0's grouped to obtain the POS expression $Q+ = \overline{R}(S + Q)$. These two equations describe direct gate implementations (NOR or NAND). The NOR implementation is shown in Figure 10.6a and the NAND implementation is shown in Figure 10.6b. To derive the NOR implementation, expand the POS expression using De Morgan's law to convert the expression into a series of NOR operations:

$$Q+ = \overline{R}(S + Q) = \overline{\overline{\overline{R} + \overline{(S + Q)}}} = \overline{R + \overline{(S + Q)}}$$

In Figure 10.6a, the $\overline{S + Q}$ expression is implemented by G1, while the $R + \overline{(S + Q)}$ expression is implemented by G2. To derive the NAND implementation, expand the SOP expression using De Morgan's law to convert the expression into a series of NAND operations:

$$Q+ = S + \overline{R}Q = \overline{\overline{S}(\overline{\overline{R}Q})}$$

In Figure 10.6, the $\overline{R}Q$ expression is implemented by G1. Notice that an inverter is used to obtain the complement of R. The $\overline{\overline{S}(\overline{R}Q)}$ expression is implemented by G2 with an inverter used to obtain the complement of S. With the S-R latch schematic derived, the timing diagrams will be shown.

10.1.4.1 S-R latch timing diagram

The timing diagram in Figure 10.7a shows different S and R input combinations triggering different operations as previously discussed. Before

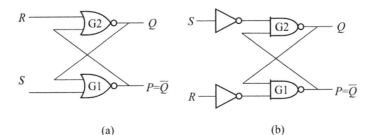

(a)   (b)

Figure 10.6   S-R latch implementations: (a) NOR gate based and (b) NAND gate based.

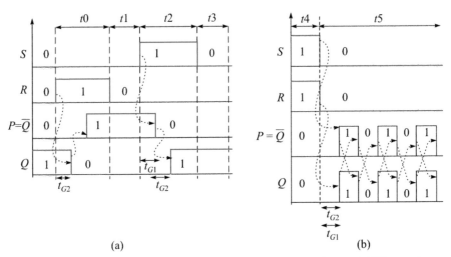

(a)   (b)

Figure 10.7   S-R latch timing diagram. (a) Timing for HOLD, RESET and SET operations (valid inputs) and (b) timing showing resulting oscillations due to invalid input $R = S = 1$ switching to $R = S = 0$.

time t_0, S and R are assumed to be 0; therefore, the S-R latch is performing the HOLD operation. Assuming the latch was storing 1 initially, output Q remains at 1 and output $\overline{Q}$ stays at 0. During t_0, R goes HIGH and S stays LOW, performing the RESET operation. The output Q becomes 0 after a gate delay, t_{G2}, associated with G2 in Figure 10.6a. The output $\overline{Q}$ changes to 1 after the change in Q propagates through G1. t_{G1} is the delay associated with G1. The change in R is therefore realized in Q after a delay of t_{G2}. If the outputs of both Q and $\overline{Q}$ are needed, then we would need to wait for $t_{G2} + t_{G1}$ for $\overline{Q}$ to be ready.

At the beginning of t_1, input signal R changes from 1 to 0, while input S stays at 0. Therefore, the latch performs the HOLD operation, so Q and $\overline{Q}$

remain unchanged. At the beginning of time t_2, input S becomes 1, while R stays at 0. The latch now performs the SET operation, so Q changes to 1 and $\overline{Q}$ changes to 0. The change in $\overline{Q}$ is observed after a gate delay of t_{G1}, and the change in Q is observed after a total delay of $t_{G2} + t_{G1}$ due to the gate delays of G2 and G1. At the beginning of time period t_3, S switches to 0, signifying the HOLD operation; hence the outputs Q and $\overline{Q}$ retain their previous values. Figure 10.7b shows what happens when the S-R latch switches from the invalid state $R = S = 1$ to the HOLD state, $R = S = 0$. If $t_{G2} = t_{G1}$, then the oscillation that occurs will go on indefinitely. The next section will explain the oscillating behavior in more detail.

10.1.4.2 Problems with the S-R latch

There are two problems associated with the S-R flip-flop (FF) implementations (NAND or NOR gate based). In Figure 10.7, the invalid input combination, $R = 1$ and $S = 1$, is shown for NAND gate-based implementation. This combination is invalid because both outputs Q and $\overline{Q}$ will switch to 1. This behavior is a problem because the relationship between Q and $\overline{Q}$ is reduced to a non sequitur: $\overline{Q}$ means the complement of Q, so how can both be equal?

The second problem with this input combination is explicated with Figure 10.8. Starting from the invalid state, R and S both switch from 1 to 0 simultaneously. Under this circumstance, there is an uncertainty in the final values of the outputs Q and $\overline{Q}$. When the inputs switch to 0, the inverters evaluate to 1 and the NAND gates evaluate to 0. The switching of Q and $\overline{Q}$ to 0 will occur after a NAND gate delay, and since these outputs are inputs to the alternative NAND gates, the NAND gates will again switch Q and $\overline{Q}$ to 1 after a gate delay. This interchanging of 0's and 1's at the output will continue to occur because the NAND gates will continually evaluate the changing signals at their inputs.

If the gate delays of both NAND gates are identical, then the oscillation continues forever; but if the gate delays are unequal, then the outputs will

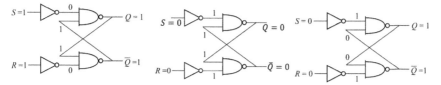

Figure 10.8 Graphical depiction of the race condition associated with the S-R latch.

converge to an indeterminate state, either 1 or 0. The convergence of the output due to unequal delays is known as the *race condition*.

10.2 Clocked Flip-Flops

In large sequential networks, it is common practice to synchronize the operation of all FFs by a common clock or pulse generator. A clock is defined as a periodical train of waveforms, as shown in Figure 10.9. The waveform is at high level for an interval of time T_1 and at low level for duration T_2. The period of the clock, T is equal to $T_1 + T_2$. The duty ratio of a clock is defined as the ratio of the time period when the clock is high, T_1, to the total clock period, T. The typical duty ratio is $\leq 50\%$. There are two types of clock synchronization, level-sensitive clocking and edge-triggered clocking.

$$\text{Clock period} = T = T1 + T2$$

$$\text{Duty ratio} = \frac{T1}{T1 + T2}$$

In *level-sensitive clocking*, the output of a FF is allowed to change only during T_1, as identified in Figure 10.10. During the time duration T_1, the inputs are sampled; if the inputs change while the clock is in the T_1 phase, the output changes. During T_2, the output remains unchanged even if inputs change. As a design rule, to avoid changing inputs immediately translated to changing outputs; circuit designers make sure inputs remain unchanged during T_1. This scheme enables the output to change only once during T_1. In summary, inputs change during T_2 and outputs change during T_1; and equally, inputs remain unchanged during T_1 and outputs remain unchanged during T_2.

In *edge-triggered clocking*, the output changes only when the clock changes state from 0 to 1 (or, 1 to 0). In an edge-triggered FF, if the output

Clock Train (Ck)

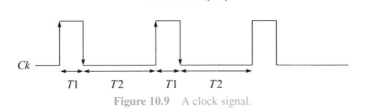

Figure 10.9   A clock signal.

Level Sensitive Clocking

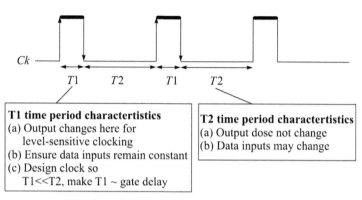

T1 time period charactertistics	T2 time period charactertistics
(a) Output changes here for level-sensitive clocking (b) Ensure data inputs remain constant (c) Design clock so T1<<T2, make T1 ~ gate delay	(a) Output dose not change (b) Data inputs may change

Figure 10.10   Level-sensitive clocking.

changes when the clock changes from 0 to 1, the FF is called a positive (+ve)-edge-triggered FF (shown with an up-pointing arrow in Figure 10.9). If the output changes when clock changes from 1 to 0, the FF is called a negative (−ve)-edge-triggered FF (shown with a down-pointing arrow in Figure 10.9).

10.2.1 Clocked S-R flip-flop

To obtain a clocked S-R FF, a clock signal is added to the S-R latch. In the NAND implementation of Figure 10.6b, the clock signal is added by changing the inverters to NAND gates, as shown in Figure 10.11.

The truth table for the obtained circuitry is provided in Figure 10.12. The table is identical to that of the S-R latch for the entries when clock (Ck) is high. When Ck is low, the outputs of the FF $(Q, \overline{Q})$ do not change no matter what the input (S, R) values are.

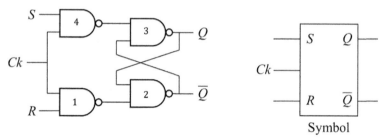

Figure 10.11   Clocked S-R flip-flop implementation and symbol.

Ck	S	R	Q+	Q̄+	Mode
0	X	X	Q	Q̄	No Action
⎍	0	0	Q	Q̄	HOLD
⎍	0	1	0	1	RESET
⎍	1	0	1	0	SET
⎍	1	1	1	1	Invalid

Figure 10.12   Clocked S-R flip-flop truth table.

The S-R FF operation is reinforced with the timing diagram in Figure 10.13. Before the first positive edge of Ck, the input S is HIGH and R is LOW; therefore the FF is configured to perform the SET operation. The output Q changes from 0 to 1 with a finite delay after Ck goes HIGH. The finite delay is equivalent to the propagation delay associated with two NAND gates of the S-R FF (Figure 10.11). Figure 10.13 also shows the FF operation during RESET (S is 0 and R is 1). Q changes after three NAND gate propagation delays after Ck goes HIGH.

Figure 10.13also shows time durations for setup and hold times, t_S and t_H, respectively. *Setup time* is defined as the duration of time the input must remain constant (stable) before the clock is asserted. *Hold time* is defined as the time required for the input to be constant (stable) after the clock edge. When designing circuits, meeting setup and hold time requirements is critical for correct operation. Setup and hold times both contribute to the determination of the maximum speed of a clocked circuit.

10.2.2 Clocked J-K flip-flop

As mentioned in the previous chapter and reiterated here, the J-K FF is a superset of the S-R FF and accommodates for the invalid state of the S-R FF by toggling. Hence, when $J = 1$ and $K = 1$, the J-K FF toggles between 0 and 1. The schematic representation of the J-K FF (Figure 10.14) is similar to that of the S-R FF (Figure 10.11). The differences introduced in the J-K schematic are use of three-input NAND gates at the inputs and the outputs Q and $\bar{Q}$ being fed back to the input gates.

Figure 10.15 presents the truth table of the J-K FF. The truth table looks exactly like that of the S-R FF, with the exception that the last case is

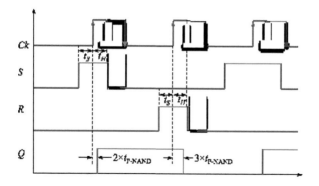

Figure 10.13 Clocked S-R flip-flop timing diagram.

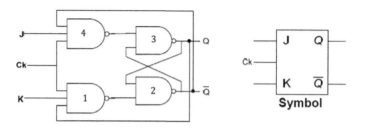

Figure 10.14 Clocked J-K flip-flop implementation and symbol.

not an invalid operation but a TOGGLE operation. As seen in the S-R FF, for any action to occur, the clock input *Ck* must be high. This is another example of level-sensitive clocking; hence the S-R and J-K FFs in the shown configurations are also referred to as *level-sensitive latches* (LSLs). In the TOGGLE operation, the output Q is set to its complement, that is, if Q was 1 it becomes 0, and if Q was 0, it becomes 1.

10.2.2.1 J-K flip-flop timing diagram

Applying a similar methodology as that used to analyze the S-R FF, the J-K FF properties are emphasized with the timing diagram in Figure 10.16. For simplification, the timing diagram does not consider accurate gate delays. Before the first clock pulse, $J = 1$ and $K = 1$, corresponding to the TOGGLE operation, so the output Q changes from 1 to 0 after some delay. At the second clock pulse, the FF performs the SET operation, since $J = 1$ and $K = 0$, so Q changes back to 1. At the third clock pulse, the FF performs the RESET operation since $J = 0$ and $K = 1$, so Q changes back to 0. At the fourth clock

Ck	S	R	Q+	$\overline{Q}$+	Mode
0	X	X	Q	$\overline{Q}$	No Action
⎍	0	0	Q	$\overline{Q}$	HOLD
⎍	0	1	0	1	RESET
⎍	1	0	1	0	SET
⎍	1	1	1	1	TOGGLE

Figure 10.15   Clocked J-K flip-flop truth table defining different FF operations.

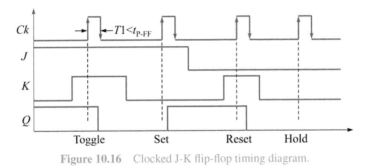

Figure 10.16   Clocked J-K flip-flop timing diagram.

pulse, $J = 0$ and $K = 0$, corresponding to the HOLD operation, so Q remains unchanged.

A vital constraint for the width of the clock pulse is that the pulse width T_1 must be less than the FF delay t_{P-FF}. This constraint ensures that in TOGGLE mode, the J-K FF changes only once in one clock pulse. If t_{P-FF} is less than T_1, the output can toggle more than once, depending on the specific value of the FF delay. Repercussions to violating this clock constraint are discussed next.

10.2.2.2 Problems with the J-K flip-flop

As previously mentioned, in TOGGLE mode, if $t_{P-FF} < T_1$, then the J-K FF output may toggle more than once, depending on the value of the FF propagation delay in relation to T_1. This operating condition is known as the *race-around condition*, since the change in output becomes the input that causes another change in output. The equivalent circuit for this condition is shown in Figure 10.17a, where the three-input NAND gates essentially

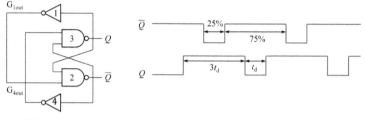

Figure 10.17   Trace of the J-K FF in the race condition for $5 \times t_d$.

operate as inverters, since *Ck, J,* and *K* signals are 1. In this condition, only the current outputs Q and $\overline{Q}$ and the NAND gate delay (t_d) determine the next state values for Q and $\overline{Q}$. If $T_1 > t_d$, then Q is indeterminate, for it may be either 0 or 1, depending on number of toggles (a function of t_d) within the clock pulse duration (T_1).

When the circuit in Figure 10.17a is examined, it can be seen that there are two ring oscillator paths. The first path is through the gates $1 \rightarrow 2 \rightarrow 3 \rightarrow 1$, and the second path is through the gates $4 \rightarrow 3 \rightarrow 2 \rightarrow 4$. Figure 10.17b shows the output waveform of a race condition when $T_1 \gg t_d$. In this waveform, after the output Q toggles for the first time to 1, G_{1out} changes to 0, making $\overline{Q}$ change to 1. If the delays of all four gates (remember gates 1 and 4 are actually NAND gates operating as inverters) are equal, gate 1's output G_{1out} changes after t_d and gate 2's output $\overline{Q}$ changes after $2 \times t_d$. At this point, both inputs of gate 3 are 1 and will remain so until G_{4out} becomes 0 due to $\overline{Q}$ changing to 1. Following this analysis will yield the table in Figure 10.18 and also the waveform for Q and $\overline{Q}$ in Figure 10.17b. The bold face values in Figure 10.18 indicate which values have changed from the previous time step. The waveforms indicate that the signal Q remains toggled for one gate delay (t_d) and then toggles back to the original value and remains in that state for three gate delays.

10.2.3 Solutions to the race-around problem

The race-around problem is avoided using two different methods. The first is by narrowing the clock pulse width, and the second is by separating the inputs of the FF from the outputs. Both methods will be explained in more detail in this section.

10.2.3.1 Narrowing the clock pulse width

This method of overcoming the race-around problem is also known as edge triggering. To reduce T_1, the circuit schematic in Figure 10.19 is used. The pulse-narrowing circuitry is composed of an inverter with delay t_{P1} and an AND gate with delay t_{P2}.

Figure 10.20 provides a timing diagram of the pulse-narrowing circuitry starting at an initial state where the clock signal A is LOW, the inverter output B is HIGH, and the modified clock signal C is LOW. When A goes HIGH, B goes LOW after an inverter delay of t_{P1}. During this delay interval, both nodes A and B (inputs to the AND gate) are HIGH. Therefore, the AND gate output C will evaluate to HIGH after a delay of t_{P2}. C eventually goes LOW when B evaluates to LOW, making one of the AND gate inputs LOW. This method always provides a modified clock signal C with pulse width t_{P1} that is much smaller than the original clock pulse width T_1. This method guarantees the J-K FF condition that $T_1 < t_d$. This is because the delay of an inverter is less than the delay of a FF.

The sample circuitry in Figure 10.19 implements positive edge triggering since the change in A from LOW to HIGH triggers the creation of the narrow pulse. If instead, the HIGH-to-LOW transition causes a narrowed pulse, then the pulse-narrowing circuit would implement negative edge triggering. Figure 10.21 shows the negative and positive edges of the clock signal used for edge

t	Q	G_{1out}	$\overline{Q}$	G_{4out}
1	1	1	0	1
t_d	1	0	0	1
$2*t_d$	1	0	1	0
$3*t_d$	0	0	1	0
$4*t_d$	1	1	1	0
$5*t_d$	1	1	0	0

Figure 10.18   Trace of the J-K FF in the race condition for $5 \times t_d$.

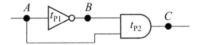

Figure 10.19   Clock pulse–narrowing circuitry used to realize edge triggering.

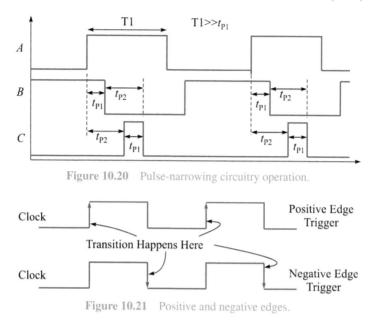

Figure 10.20   Pulse-narrowing circuitry operation.

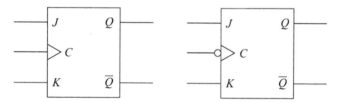

Figure 10.21   Positive and negative edges.

triggering, and Figure 10.22 shows the symbols for positive and negative edge–triggered J-K FFs.

Aside: Use of the pulse-narrowing concept in an actual CMOS implementation
The previous discussion presented a concept of how edge triggering can be realized. Actual implementations may look different, but the concept remains

Figure 10.22   J-K FF symbols. (Left) Positive edge triggered. (Right) Negative edge triggered.

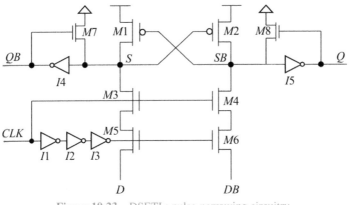

Figure 10.23   DSETL: pulse-narrowing circuitry.

the same. A variant of a D FF implementation, the dual-rail static edge-triggered latch (DSETL), is shown in Figure 10.23. The DSETL[2] evidently provides superior performance in power delay product, layout area, and speed. The three-inverter chain in Figure 10.23 produces the complement of *CLK* used to generate a smaller pulse that allows *D* and *DB* to pass for a short amount of time through M3 and M5, and M4 and M6, respectively. This clock-narrowing scheme uses the same concept, except instead of a one-inverter delay for narrowing the clock pulse; the new pulse width is equal to a three-inverter delay. A sample waveform that shows the DSETL takes 175 ps to propagate the data from *D* to *Q* is given in Figure 10.24. The power consumption comparison with other technologies (see the paper at footnote to learn the competing technologies referenced in the chart) is also shown in Figure 10.24.

You may wonder whether we can build flip-flops that transfer data at both edges of the clock to improve the system performance. The answer is Yes.[3]

---

[2]L. Ding, P. Mazumder and N. Srinivas, "A Low Power Dual-line Static Edge Triggered Flip-flop," Proc. of the IEEE International Symposium on Circuits and Systems, Sydney, May 2001, pp. 645–648.

[3]Most videogame aficionados are familiar with memory (DRAM) modules that they often add to their computers to improve the speed of the games. Many of these DRAM memories transfer data at both positive and negative edges of the system clock to double the data rate (DDR). Such memories are called the double data rate synchronous dynamic random-access memory (DDR SDRAM). Chapter 21 describes computer memories that include DRAM, which is used as main memory in the majority of computers.

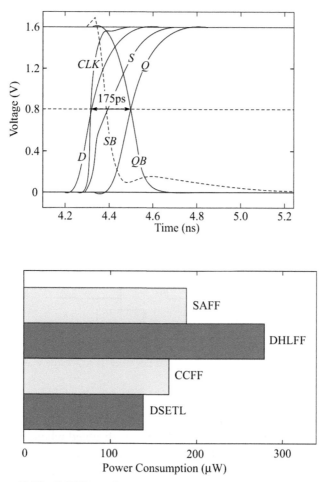

Figure 10.24 DSETL: performance and power consumption characteristics.

10.2.3.2 Separating the inputs and outputs of the flip-flop

The second method to avoid the race-around condition in the J-K FF is to separate the FF's inputs and outputs. This method involves using two FFs and is explicated first using D LSLs in Figure 10.25. Remember, an LSL is essentially a FF that utilizes level-sensitive clocking.

10.2.3.2.1 Master–Slave D Flip-Flop

In the configuration in Figure 10.25, if the master D LSL is clocked with Φ, then the slave LSL must be clocked with $\overline{\Phi}$, the complement of Φ. This

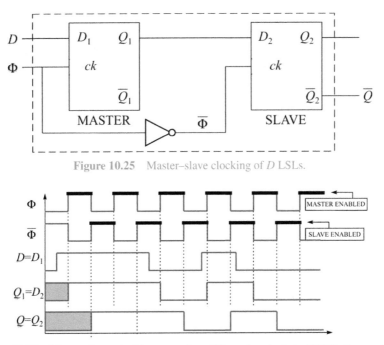

Figure 10.25   Master–slave clocking of D LSLs.

Figure 10.26   Master–slave clocking operation. When the clock is HIGH, the master is enabled and the slave is in the HOLD state. When the clock is LOW, the master is in the HOLD state and the slave is enabled.

clocking configuration is known as *master–slave clocking*. By inverting the clock of the slave D LSL, the output evaluation is disabled when the input is enabled, and vice versa. Two LSLs work essentially as one storage unit; the input D is transferred to the output Q at most once during a clock period, and hence the race-around condition would be eliminated in a master–slave J-K FF, as will be shown later.

Figure 10.26 explains the master–slave clocking operation of the schematic in Figure 10.25. When the clock (Φ) goes HIGH, the input to the master LSL is sampled and the master output immediately changes to the sampled value. When Φ goes LOW, $\overline{\Phi}$ goes HIGH, causing the slave LSL to sample the output of the master LSL—the output of the slave LSL immediately changes to the sampled value of the master LSL. Even though the change in output is observed at the negative edge of the clock, the FF is not negative edge triggered. Master–slave clocking is a form of pulse-triggered-based clocking by IEEE standards.

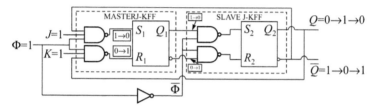

Figure 10.27 Master–slave clocking of J-K FFs illustrating no race-around condition when $J = K = 1$ and clock is HIGH.

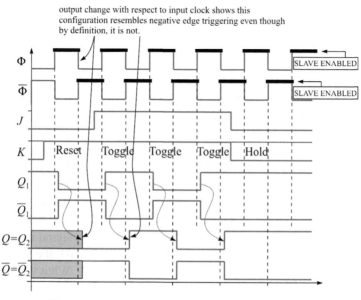

Figure 10.28 Master–slave clocked J-K FF operation.

10.2.3.2.2 Master–Slave J-K Flip-Flop

Compared to the master–slave D FF, the master–slave configuration for the J-K FF looks a bit different (Figure 10.27), but the underlying concept is exactly the same. For the J-K master–slave FF, the outputs of the slave FF are fed back to the inputs of the master FF for TOGGLE mode and complementary clocks are used.

Figure 10.28 uses a timing diagram to show that the TOGGLE mode of the master-slave J-K FF does not have a race condition associated with it. Initially, a RESET operation is performed to set both master and slave FF outputs (Q_1 and Q_2) to LOW. Afterward, inputs J and K are made HIGH before the positive clock phase signaling a TOGGLE operation. At the second

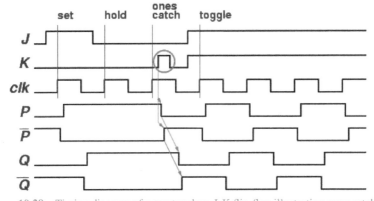

Figure 10.29   Timing diagram of a master slave J-K flip-flop illustrating ones catching.

clock pulse, the master FF performs the TOGGLE operation setting Q_1 to HIGH. When the clock goes LOW, its complement $\overline{\Phi}$ goes HIGH. Since Q_1 is HIGH and $\overline{Q_1}$ is LOW, the slave FF performs the SET operation— the output Q_2 goes HIGH. During one clock transition the output can only change once because when the clock is LOW, the master FF performs the HOLD operation; thus the inputs to the slave FF are stable. Therefore, there is no race-around condition during the TOGGLE operation. Going back to the timing diagram, another TOGGLE operation is performed since both J and K are still HIGH, so Q_1 toggles to LOW when the clock signal is HIGH. When the clock goes LOW, since Q_1 is LOW and $\overline{Q_1}$ is HIGH, the slave FF performs the RESET operation and the output Q_2 goes LOW. What we notice in this scheme is that the master FF can go through the various schemes SET, RESET, HOLD, and TOGGLE. Since the slave always gets complementary inputs from the master FF, it can only perform SET, RESET, and HOLD. This clocking scheme shows that the FF output Q changes at the negative edge of clock, but remember, by definition, this is not negative edge triggering but pulse edge triggering.

   Figure 10.29 shows a timing diagram for the master–slave J-K flip-flop. Notice that outputs Q and Q' change after the falling transition of the clock signal, *clk*. Also notice that the waveforms of the internal signals (output of master flip-flop) P and P' are similar to signals Q and Q' (with a delay of half clock cycle).

   An important phenomenon in master–slave J-K flip-flops is *ones catching*. If inputs J or K are 1 at any point during the active phase of the clock cycle, then they are effectively interpreted as a 1 by the flip-flop circuit. The reason for the ones catching problem can be explained as follows. The master latch is enabled during the active phase of the clock. Its state can thus be changed by an active pulse in J or K when *clk* is high. When this happens, the state of the master latch is kept even if the input goes low again—the master *catches* the 1. The slave latch responds to the new state of the master when the clock signal goes low. If the flip-flop is implemented using NAND gates, the circuit is prone to the *zeros catching* problem.

Aside: Master–Slave D FF implementation in commercial CMOS technology
Since the Master–Slave D and J-K FFs discussed here require too many transtors to implement their functionalities, in CMOS technology they are designed economically with fewer transistors in order to accrue superior speed and reduced power consumption.

Figure 10.30 shows a typical master-slave D flip-flop implemented using CMOS technology. Essentially, both master and slave sections consist of basic storage unit comprising a pair of back-to-back connected inverters, which will require $4 + 4 = 8$ transistors altogether for both the sections. Additionally, 4 NMOS transistors are used in this diagram to ensure that when the master is enabled, the slave will be disabled, and vice versa. Further, when the master is enabled by asserting CLK = 1, the feedback between inverter pair is broken as the NMOS transistor Q2 activated by $\overline{\text{CLK}}$ will turn OFF. However, the slave will be configured into storage mode as the NMOS transistor, Q4 in the feedback path will turn ON connecting the inverter-pair in back-to-back configuration. Also, note that when CLK = 1, Q1 is turned ON and Q3 is turned OFF, thereby ensuring that slave is in disabled mode, while the D input (data) is sampled by the master at node X as $\overline{D}$. When CLK = 0, i.e., $\overline{\text{CLK}} = 1$, the output of the slave Q will be set to $\overline{X} = \overline{\overline{D}} = D$, thereby transferring the data (D) to Q at the negative edge of the clock. Also, when $\overline{\text{CLK}} = 1$, the NMOS transistor, Q2 in the feedback path of inverter pair in the master section will turn on, holding the sampled value of D at X when CLK = 1, i.e., the master was enabled[4].

Aside: Edge-triggered FF implementation (7474)
Figure 10.31 shows a positive-edge-triggered commercial D FF (7474) schematic and truth table. The number 7474 defining this FF circuitry is fixed and only refers to this circuit configuration. This number system

---

[4]In order to improve the performance of a D FF, in practice a PMOS transistor is added in parallel to each NMOS transistor (i.e., Q1, Q2, Q3 and Q4), and the augmented parallel PMOS+NMOS switch is called Transmission Gate (TG).

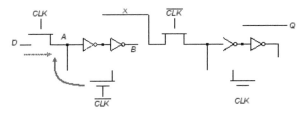

Figure 10.30   Efficient implementation of Master–Slave D FF in CMOS technology

allows for standards to be set, so no matter which technology (e.g., complementary metal-oxide semiconductor [CMOS], transistor–transistor logic [TTL], emitter-coupled logic [ECL]) a company uses to fabricate its FF, it ensures compatibility and functionality.

In the logic diagram, the inputs to the FF ($\overline{S_D}$, $\overline{C_D}$, CP, and D) are used to determine the outputs (Q and $\overline{Q}$). $\overline{S_D}$ and $\overline{C_D}$, corresponding to PRESET and CLEAR, respectively, are presented as complements, which means that they are activated when either of them is LOW. A problem arises when both are asserted LOW at the same time; this case presents an invalid condition, as shown in the truth table (Q and $\overline{Q}$ are HIGH). The truth table also hints that irrespective of the value of the clock (CP), $\overline{S_D}$ and $\overline{C_D}$ may change the outputs. $\overline{S_D}$ and $\overline{C_D}$ are therefore asynchronous inputs because their influence on Q and $\overline{Q}$ is not coordinated by CP. When $\overline{S_D}$ and $\overline{C_D}$ are deactivated, CP synchronizes the transfer of D to the outputs. Whenever CP is LOW, the FF holds its current state. This FF also shows that in industry, clock inversion is usually avoided because of various challenges it may pose due to inverter delay.

10.2.4 IEEE symbols and flip-flop types

Due to different books and varying conventions for FFs, IEEE has standardized various symbols to signify latches, master-slave FFs, and edge-triggered FFs. Figure 10.32 lists some standard IEEE FF symbols and types. As shown, master-slave is not considered edge-triggered. When using design tools, cell libraries will contain different types of FFs.

Aside: Commercial FF with parallel inputs and parallel outputs (74175)
The IC 74175 consists of four edge-triggered D FFs with individual D inputs and Q and $\overline{Q}$ outputs. The clock (CP) and master reset ($\overline{MR}$) signals are common. The four FFs will store the state of their individual D inputs on the LOW to HIGH clock transition, causing individual Q and $\overline{Q}$ outputs to

Truth Table (Each Half)

Inputs				Outputs	
$\overline{S}_D$	$\overline{C}_D$	CP	D	Q	$\overline{Q}$
L	H	X	X	H	L
H	L	X	X	L	H
L	L	X	X	H	H
H	H	$\diagup$	H	H	L
H	H	$\diagup$	L	L	H
H	H	L	X	Q_0	$\overline{Q}_0$

H — HIGH Voltage Level
L — LOW Voltage Level
X — Immaterial
$\diagup$ — LOW-to-HIGH Clock Transition
$Q_0(\overline{Q}_0)$ — Previous $Q(\overline{Q})$ before LOW-to-HIGH Transition of

- Positive Edge Triggered
- Asynchronous Active Low
  PRESET and CLEAR
- Invalid Output (Q=Q′ = 1)
  if PRESET = CLEAR = 0

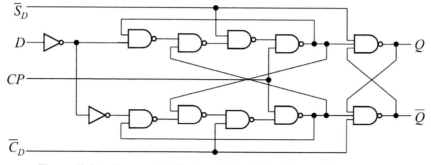

Figure 10.31 Commercial flip-flop: 7474 obtained from a TTL data book.

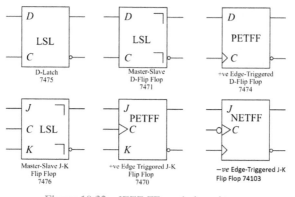

Figure 10.32 IEEE FF symbols and types.

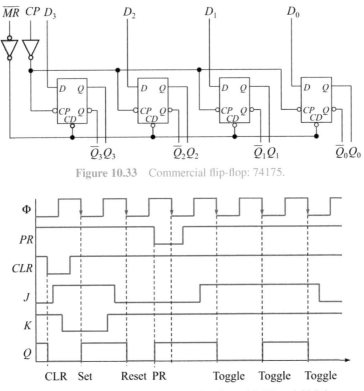

Figure 10.33   Commercial flip-flop: 74175.

Figure 10.34   Timing diagram of a J-K flip-flop with PR and CLR inputs.

follow. A low input on $\overline{MR}$ will force all Q outputs low and $\overline{Q}$ outputs high independent of *CP* or *D* inputs. The 74175 is useful for general logic applications where a common master Reset and clock are acceptable. The 4-bit FF allows parallel data transfer from $\{D_0, D_1, D_2, D_3\}$ to $\{Q_0, Q_1, Q_2, Q_3\}$ (Figure 10.33) and is called a Register, where data is stored in a bank of clocked flip-flops.

10.2.5 Timing analysis of flip-flop problems

10.2.5.1 Example 1: timing diagram of a J-K flip-flop

Problem: Given the negative-edge-triggered J-K FF timing diagram in Figure 10.34, explain how the output Q is obtained. PR and CLR correspond to PRESET and CLEAR, respectively.

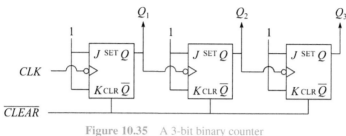

Figure 10.35 A 3-bit binary counter

Solution: The first step to analyzing this timing diagram is to mark all the negative edges of the clock. The diagram in Figure 10.32 has negative edges of CLK labeled, and these will signify the areas where synchronous inputs will affect the output.

The second step is to pay attention to the asynchronous inputs, PR and CLR. Since asynchronous inputs override clock synchronization, determine Q at the moments when Q depends on PR and CLR. Note that PR LOW sets Q HIGH, and CLR LOW sets Q LOW; these two regions are shown with markers as PR and CLR.

The third step is to determine the synchronous inputs at the marked points on the clock edge. The first marking (discounting the one that falls within asynchronous control of Q) to consider has $J = 1$ and $K = 0$, signifying SET, so Q should change to HIGH at this clock edge. The next clock edge, $J = 0$ and $K = 1$ signifies RESET, so Q should be LOW at this clock edge. The next significant clock edges have $J = K = 1$, the FF is in TOGGLE mode, so Q toggles in the three clock edges. Regions of Q's response attributed to FF inputs are identified in Figure 10.34.

10.2.5.2 Example 2: timing diagram of J-K flip-flops configured as a counter/binary divider

Problem: Given the 3-bit binary counter in Figure 10.33, determine the output waveform.

Solution: The counter is implemented with three J-K FFs with J and K inputs set to 1, that is, the J-K FFs are set in TOGGLE mode. FF #1 is clocked with CLK, FF #2 is clocked with Q_1, and FF #3 is clocked with Q_2. Each FF is connected to a common, active-LOW, asynchronous CLEAR input.

As in Example 1, the negative edges of CLK are marked; and then the asynchronous inputs are determined, so CLR set LOW will cause Q_1 through Q_3 to initialize to 0. Following along with Figure 10.36, the next step is to determine Q_1 since FF #2 depends on Q_1. At every negative CLK edge, Q_1 toggles because FF #1 is configured in TOGGLE mode. Q_1 is the clock input for FF #2 so Q_2 toggles every negative edge of Q_1. After determining Q_2, the negative edges of Q_2 are used to identify when FF #3 will toggle, i.e. determine Q_3 toggle points.

Essentially, Q_1 toggles with double the toggling frequency of Q_2, and Q_2 toggles with double the toggling frequency of Q_3. At each CLK pulse after CLR is deactivated, Q_3, Q_2, and Q_1 follow the sequence 000, 001, 010, ... 111, 000 ... Therefore, the output is a binary counting sequence.

10.2.5.3 Example 3: timing diagram of D flip-flops configured as a ring counter

Problem: Given the 4-bit ring counter in Figure 10.37, determine what the waveform looks like if the circuit is first initialized.

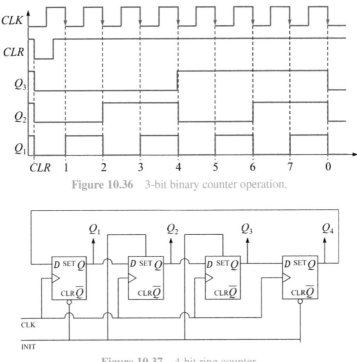

Figure 10.36   3-bit binary counter operation.

Figure 10.37   4-bit ring counter.

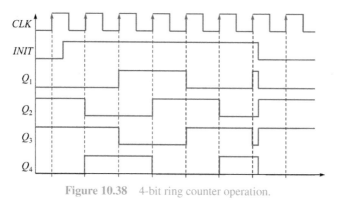

Figure 10.38   4-bit ring counter operation.

Solution: The 4-bit ring counter is implemented using four positive-edge-triggered D FFs. From left to right, the output of one FF is connected to the input of the next FF. The output of FF #4 is the input to FF #1. The FFs are clocked with the same clock signal, C_k. The CLR input of the first and fourth FFs and the *SET* inputs of the middle two FFs are connected to the *INIT* signal.

Figure 10.36 provides the solution graphically. When the asynchronous input *INIT* is LOW, the RESET operation sets $Q_1 = Q_4 = 0$ and the PRESET operation sets $Q_2 = Q_3 = 1$. Therefore the outputs Q_1, Q_2, Q_3, Q_4 are initialized to 0110. When the *INIT* signal goes HIGH, the counter observes clock synchronization. In this mode, the value stored in each D FF is passed to the next FF it is connected to. This operation causes the output bits to shift right, and the last bit (Q_4) is moved to the first position (Q_1). Hence, at each positive clock edge when the *INIT* signal is HIGH, a shifting operation is performed. Since the outputs are initialized to 0110, the sequence is as follows: 0011, 1001, 1100, 0110, 0011 . . . When the *INIT* signal goes low, the counter stops shifting and sets the outputs to 0011.

10.2.6 Effect of the width of the set pulse

As previously mentioned, the hold time and setup time for FFs are very important metrics that if violated could yield incorrect circuit operation. The effect of setup time violation and the reason why a circuit could perform in an unexpected way are explained with the aid of Figure 10.39.

Figure 10.39 shows three *S* pulses of differing widths and their effects on S-R FF performance. Given $S = 0$, $R = 0$, $Q = 0$, and $Q' = 0$, the goal for all three cases is to set Q to 1 by setting S to 1. In the first case, when the *S* pulse is wide enough (setup time is observed), then correct behavior follows and *Q* is set to 1. This case is shown in the output *Q–Q'* curve as the high voltage point. In the third case, when the *S* pulse is too short, the FF does not respond, and this case is shown as the low voltage point in the *Q–Q'* curve. This behavior is sometimes desired because the *S* input may be noisy, and usually noise presents itself as short-duration pulses. Noise immunity is desired in circuit design. In the middle case, when the *S* pulse width is almost enough to toggle the gate output, the output voltage gets stuck at an intermediate voltage value, as shown in the *Q–Q'* curve. This state is called the *metastable* state. The metastable state is not unstable, since *Q* does not oscillate. But metastability is still undesired behavior for output logic level (HIGH or LOW) is uncertain.

10.2.7 Dealing with meta stability and asynchronous inputs

Meta stability is tackled in the same way as *asynchronous inputs* such as CLEAR and PRESET inputs of a 7474 flip-flop. These asynchronous inputs are independent of the synchronous clock signal and their changes immediately effect the operation of the clocked sequential network. When asynchronous inputs are handled by logic, special care must be given in designing such logic, since any glitches in the asynchronous input can potentially cause a flip-flop to be cleared or preset incorrectly. In order to avoid any malfunction due to inadvertent glitches, a synchronizer flip-flop is used so that instead of connecting the external signal to several storage

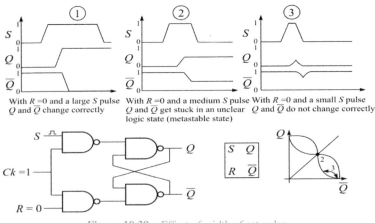

With *R* =0 and a large *S* pulse   With *R* =0 and a medium *S* pulse   With *R* =0 and a small *S* pulse
Q and *Q̄* change correctly   *Q* and *Q̄* get stuck in an unclear   *Q* and *Q̄* do not change correctly
                                      logic state (metastable state)

Figure 10.39   Effect of width of set pulse.

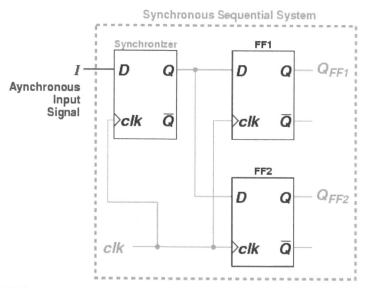

Figure 10.40 At the asynchronous input a synchronizing D flip-flop is used to overcome meta stability and timing constraints.

elements, the signal should first be fed into a synchronizer flip-flop, as shown in Figure 10.YY. Feeding the input signal through a flip-flop guarantees that, inside the circuit, signal changes will occur only as the result of clock signal events.

If an asynchronous signal is fed to the input D of a flip-flop, then there is a possibility that the setup or hold timing constraints of the storage device will be violated. If the asynchronous signal is connected to several flip-flops instead of just one, then the likelihood of timing constraint violations increases significantly.

Figure 10.40 is also used to overcome the effects of meta stability and will be deployed in the design of a true random number generator (TRNG) in Chapter 12. In some cases, a series of synchronizer flip-flops are needed to eliminate the anomalous effects of meta stability.

10.3 Conclusion and Key Points

1. There are primarily two types of flip-flops: S-R and J-K flip-flops. Both T and D flip-flops are derived from J-K flip-flops.

2. J-K flip-flop suffers from race condition when $J = K = 1$ and clock is High for longer than the gate propagation delays, thereby causing the output to toggle continuously as long as clock is high and both J and K are high.
3. When $J = K = 1$ and clock is High, the J-K flip-flop degenerates into two oscillators that cause Q and Q' to toggle continuously, and at sometimes $Q = Q'$, i.e. their output values overlap.
4. S-R flip-flop can also be made to oscillate if at first $S = R = 1$ is asserted simultaneously and then the flip-flop is put into Hold mode by simultaneously holding $S = R = 0$. In this case, the S-R flip-flop breaks into oscillation with $Q = Q'$ during the time of oscillation.
5. Flip-flop mainly employs three types of clocking: 1. Level-sensitive, 2. Edge-triggered, and 3. Pulse triggering or Master–Slave clocking.
6. Depending on the clocking scheme, the input signals at the input pins of a flip-flop are sampled at different window of time.
7. The concept of set-up and hold times for a flip-flop are very important since the flip-flop may settle at the meta-stable state where $Q = Q'$ mainly because of violation of set-up and hold times. The output values of a flip-flop must become stable before input signals can change.
8. Flip-flops may contain asynchronous CLEAR and PRESET inputs that override the clock and other inputs. If PRESET is asserted, $Q = 1$ and $Q' = 0$. If CLEAR is asserted, $Q = 0$ and $Q' = 1$.

10.4 Problems on Flip-Flops

1. You are given the circuit below and are asked to analyze it. Is this circuit a useful latch? If so, explain its operation. If not, what would you change it to make it a useful latch?

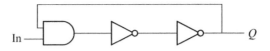

2. For the *MN* latch given below, identify its operations on its truth table and provide its characteristic equation.

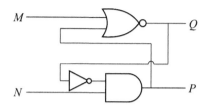

3. Is there a problem with the *MN* latch provided above? If so, provide the conditions under which the problem occurs.
4. For a clocked S-RFF, the timing diagram below is given. Complete the diagram based on the input waveforms provided.

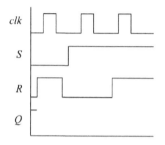

5. Given two edge-triggered D FFs, find the values of Q_2.

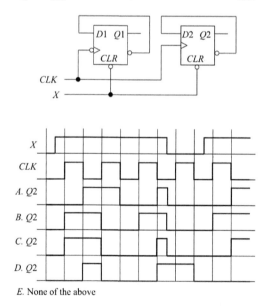

6. Given two edge-triggered J-K FFs, complete the timing diagram below.

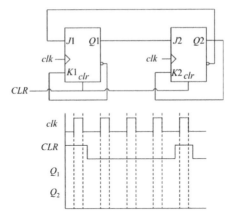

7. Given the following sequential circuit, find the values of Q_1, Q_2, and Q_3, at time t_2.

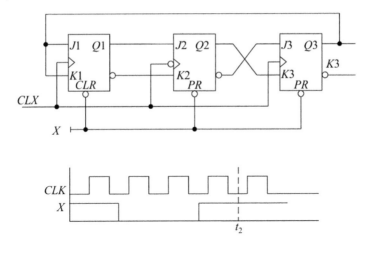

At time t_2
A. $Q_1 \ Q_2 \ Q_3 = 010$
B. $Q_1 \ Q_2 \ Q_3 = 001$
C. $Q_1 \ Q_2 \ Q_3 = 011$
D. $Q_1 \ Q_2 \ Q_3 = 000$
E. None of the above

8. This figure shows a new type of FF, the ABD FF. Which of the following is its characteristic equation?

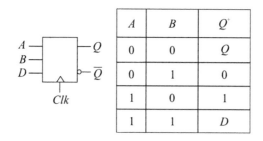

A	B	Q'
0	0	Q
0	1	0
1	0	1
1	1	D

(a) $Q^+ = A'B + BQ' + ABD$
(b) $Q^+ = AB' + B'Q + AD$
(c) $Q^+ = AD + B'Q$
(d) $Q^+ = AB + DQ$
(e) $Q^+ = A'D + B'Q$

9. Modify the binary counter below such that it only counts up to 6 instead of 7.

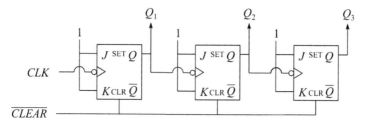

10. You are given three positive-edge-triggered D FFs connected as shown below. Complete the timing diagram with the waveforms of Q_1, Q_2, and Q_3.

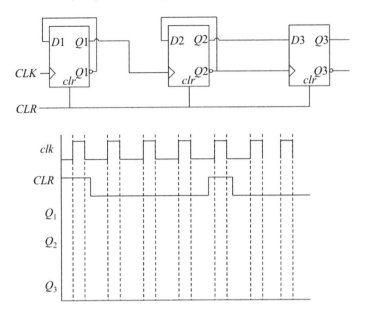

11. Analyze the operation of the following edge-triggered D flip-flop and then write Verilog programs to produce the simulation results at the maser and slave outputs. Validate your manual analysis by comparing with simulation results.

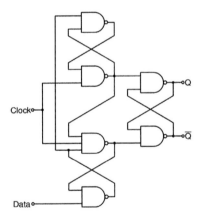

12. Analyze the operation of the following edge-triggered D flip-flop and then write Verilog programs to produce the simulation results at the maser and slave outputs. Validate your manual analysis by comparing with simulation results.

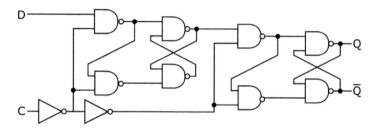

13. Analyze the operation of the following master–slave J-K flip-flop configured in the form of D FF and then write Verilog programs to produce the simulation results at the maser and slave outputs. Validate your manual analysis by comparing with simulation results.

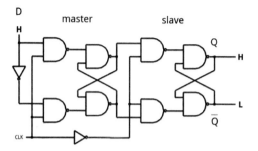

14. Analyze the operation of the following edge-triggered D flip-flop and then write Verilog programs to produce the simulation results at the maser and slave outputs. Validate your manual analysis by comparing with simulation results.

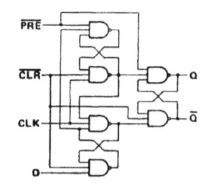

15. Write a Verilog program for the given edge-triggered flip-flop as shown in the figure. You must produce the simulation results along with your Verilog code.

The ACT109 consists of two high-speed completely independent transition clocked JK flip-flops. The clocking operation is independent of rise and fall times of the clock waveform.

LOW input to SD (Set) sets Q to HIGH level

LOW input to CD (Clear) sets Q to LOW level

Clear and Set are independent of clock

Simultaneous LOW on CD and SD makes both Q and Q HIGH

Truth Table (each half)

Inputs					Outputs	
S_D	C_D	CP	J	K	Q	Q̄
L	H	X	X	X	H	L
H	L	X	X	X	L	H
L	L	X	X	X	H	H
H	H	L	L	L	L	H
H	H	L	H	L	Toggle	
H	H	L	L	H	Q_0	$Q̄_0$
H	H	L	H	H	H	L
H	H	L	X	X	Q_0	$Q̄_0$

H = HIGH Voltage Level
L = LOW Voltage Level
L = LOW-to-HIGH Transition
X = Immaterial
$Q_0(Q̄_0)$ = Previous $Q_0(Q̄_0)$ below LOW-to-HIGH Transition of Clock

Logic Diagram (one half shown)

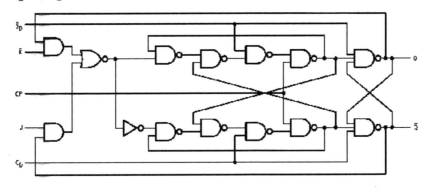

Index

243

About the Authors

Pinaki Mazumder is currently a Professor with the Department of Electrical Engineering and Computer Science, University of Michigan (UM), Ann Arbor. He was for six years with industrial R&D centers that included AT&T Bell Laboratories, where in 1985, he started the CONES Project the first C modeling-based very large scale integration (VLSI) synthesis tool at Indias premier electronics company, Bharat Electronics, Ltd., India, where he had developed several high-speed and high-voltage analog integrated circuits intended for consumer electronics products. He is the author or coauthor of more than 350 technical papers and eight books on various aspects of VLSI research works. His current research interests include current problems in nanoscale CMOS VLSI design, computer-aided design tools, and circuit designs for emerging technologies including quantum MOS and resonant tunneling devices, semiconductor memory systems, and physical synthesis of VLSI chips. Dr. Mazumder is a Fellow of the American Association for the Advancement of Science (2008). He was a recipient of the Digitals Incentives for Excellence Award, BF Goodrich National Collegiate Invention Award, and Defense Advanced Research Projects Agency Research Excellence Award.

Idongesit E. Ebong, Ph.D. is a patent agent at Nixon Peabody in Chicago. He practices in the technical areas of electrical, computer, and mechanical engineering. He received his Ph.D. degree in electrical engineering from the University of Michigan, Ann Arbor. His research areas included digital/analog integrated circuit design, focused primarily on non-traditional devices like memristors and tunneling transistors for low power applications. Leveraging his background, he helps clients protect inventions in computer architecture and networks, digital and analog circuits, software systems, microfabrication equipment design, MEMS sensors, machine learning, and telecommunications.

9788770223614